INDIAN AGRICULTURAL ECONOMY

ABOUT THE AUTHOR

DR. R. DASGUPTA has been a faculty member of National Institute of Bank Management, an apex level training and research institution of banking industry since 1984. He had earlier worked at Institute of Economic Growth, Delhi, Perspective Planning Division of Indian Planning Commission and Centre for Development Studies, Trivandrum. He was honorary consultant to Bureau of Industrial Costs and Prices and was involved in estimation of demand for commercial vehicles. He obtained Ph.D. degree for his thesis Nutritional Planning in India from Delhi University. The same has been published in Institute of Economic Growth's series Studies in Economic Development and Planning.

Dr. Dasgupta is engaged with the redefinition of development; and formulation of appropriate strategies and policies to achieve the above.

INDIAN AGRICULTURAL ECONOMY

An Analysis

RAJARAM DASGUPTA

National Institute of Bank Management
Pune

Foreword by

C.H. HANUMANATHA RAO

Centre for Economic and Social Studies
Hyderabad

CONCEPT PUBLISHING COMPANY, NEW DELHI 110059

ISBN 81-7022-522-1

First Published 1994

Published and Printed by
Ashok Kumar Mittal
Concept Publishing Company
A/15-16, Commercial Block, Mohan Garden
NEW DELHI-110059 (India)

Lasertypeset by
Printing Express
H-13, Bali Nagar
NEW DELHI-110015

To

Chiropriya and Arkodeb

With the hope that their generation in their time will be imbued with the scientific temperament to realise the significance of investment in weaker sections

FOREWORD

The formal rural credit system in India is facing a severe crisis. The mounting overdues and losses are choking the channels of institutional credit, starving the promising and the honest farmers of much needed finance. Despite the phenomenal expansion of institutional credit for agriculture following the nationalisation of major commercial banks, only about 10 per cent of small and marginal farmers are able to borrow from institutional sources. The bulk of the benefits from institutional credit, especially from term-credit extended by the commercial banks, have accrued to the affluent farmers whose repayment performance has, on all accounts, been worse than that of the resource – starved small and marginal farmers. Similarly, the institutional sources are unable to meet the growing demand for credit from the less developed regions such as the eastern gangetic belt, where the dependence on the non-institutional sources is still quite high. Thus, we are still far from achieving the avowed objectives behind the nationalisation of major commercial banks.

The author brings out in this book the growing disparities, regional and class-wise, in agricultural performance and the failure of the institutional credit system to counter these trends. According to him, one of the main causes for slow agricultural growth and unequal distribution of income is the small size of the rural market which is perpetuated on account of the exploitation of the poor within a given market. He suggests that effective provision of agricultural credit to small farmers, tenants and agricultural labourers may break this vicious circle of exploitation and induce the cultivators to innovate and achieve higher level of production by enlarging their vision.

The author identifies three principal deficiencies of the financial institutions which are responsible for slow rate of agricultural growth and unequal sharing of benefits from growth. They are : (a) Lack of orientation to development which is manifested in the preoccupation with the given restricted market with a myopic outlook rather than trying to expand the market; (b) The banks are, therefore, risk averse and conservative in their attitudes; and (c) Urban bias and bias in favour of affluent farmers.

Chapter 4 of this book outlines the main elements of the author's alternative approach to agricultural credit. He pleads for a drastic reorientation of the outlook of banks by thoroughly restructuring the training of their personnel, so as to alter their value system and improve their understanding of the economy. So far as the policy framework of the banks is concerned, emphasis has to be, as per the guidelines of the Reserve Bank of India, "on the physical realisation of results in terms of output, productivity, income, employment and not merely on disbursement of credit". He suggests that deposits may be considered as proxy for output and employment as it would be difficult for the banker to estimate the latter. He pleads for higher weights being given to deposits of smaller people. Together with this, he argues, the recovery rate as performance indicator needs to be calculated differently using higher weights for bigger farmers as they have been found to be wilful defaulters.

According to the author, lending methodology too needs to be changed. He says, for example, that small farmers need to be given consumption loans so as to free them from the clutches of moneylenders and to turn them into potential depositors. This may appear to conflict with the objective of deposit mobilisation and higher recovery rate so long as performance continues to be judged by short-run outcomes, even though the long-term consequences of such a policy would be encouraging.

The author discusses the mechanics of credit planning as part of the People's Action Plan through the participation of the people in terms of listing out their priorities, selection of beneficiaries, discussion of their problems etc. He then shows how the National Institute of Bank Management while pursuing experiments on these lines since 1983 has already got some encouraging results indicating substantial improvements in the conditions of the weaker sections as also in the banks' performance in terms of loan recovery and deposit mobilisation.

For this model to work, the motivation as well as the capabilities of the bankers at the grass roots level have to improve significantly. The bankers at the grass roots should no longer be regarded as an underpriviledged class when compared to thier urban counterparts. The rural postings should attract the younger and the competent staff. The system of rewards too needs a change so that improved performance is associated with better remuneration.

A number of high-level committees, notably the Agricultural Credit Review Committee headed by Prof. A.M. Khusro and the Committee on the Financial System headed by Shri M. Narasimham have examined the causes underlying the present crisis faced by the rural financial institutions and have come out with a number of important recommendations for restoring the health of these institutions. Dr. R. Dasgupta, while not disagreeing with many of these recommendations and in fact significantly drawing upon the findings of the Khusro Committee, presents a fresh approach in this book which is innovative, bold and humane. His proposals for credit reform aimed at development and equity are consistent with efficiency and are also practicable.

C.H. HANUMANTHA RAO
Centre for Economic and Social Studies
Hyderabad

PREFACE

India is an agricultural economy where the bulk of production is related either to agricultural crops, or agro-based items or industrial goods meant for agricultural sector. Much more than this, three quarters of the Indian population is still dependent on agricultural production in rural areas. Agricultural sector, however, is yet to get proper attention it deserves. Today's urban world perhaps considers it as a necessary evil, and equates development with industrialization. However, the irony is that generally all the budget speeches must make a mention about the monsoon, and all optimistic proclamations depend upon it.

This book has therefore been written with the sole intention of reemphasising agriculture in its proper perspective in the development process. Large agricultural production alongwith a strong rural market base is a *sine-qua-non* for a vibrant industrial growth. In the absence of the above conditions, industrial growth in a country like India will take place in a haphazard fashion only. It is, therefore, extremely important in today's 'reform' environment to talk about agricultural or more correctly rural sector reform first. These include land reform in terms of both redistribution, consolidation and land development, formation of cooperatives and groups of small and marginal farmers and other non-farm producers, new methods of financing rural activities and creation of forward and backward linkage systems which are different from the existing unfriendly channels. For this a major reform is required to bring about the change in the entire delivery system. Unless these reforms are urgently undertaken, it is left to imagination whether the present *ad hoc* type of reform movements will usher in the expected growth or it will contribute to the increase in unemployment, poverty, undernutrition and misery on the one hand and filthy richness, mafia gangwars, financial indiscipline and lumpen activities on the other hand. What is required today is a new sense of development; a new list of priority; a new kind of methodology; and a new type of attitude. All this may sound a little Utopian. One, however, should not overlook the fact that for countries like Japan or South Korea the above types of reforms have been the main causes of success.

The present book while pursuing the above theme is divided into four chapters. The first chapter attempts to find out the performence of different states in agricultural and foodgrains output in terms of long-term growth, trends in growth and fluctuation in production. In the second chapter an attempt is made to analyse the importance of different factors in explaining the agricultural performance. More importantly the role of institutional credit has been underlined. If properly pursued formal credit can bring forth a dynamism in rural economy breaking the production relation barriers. The third chapter analyses agricultural credit from both commercial bank and cooperative sector, and tries to discover the behaviour of rural credit institutions. The last chapter, "An alternative approach to agricultural credit", tries to make suggestion for credit reform which is not necessarily in terms of interest rate structure or share of priority sector lending alone, but in terms of performance criterion, reward system, lending methodology, organizational design and attitudinal change also which can be brought about by a different kind of training system.

My first sincere thanks go to Professor C.H. Hanumantha Rao who not only consented to write the Foreword, but instilled a sense of meaningful economics in me during my formative years at the Institute of Economic Growth.

Ashok Mitra, Sudipto Mundle and Surjeet Singh read some of the draft papers and offered their comments which have modified the text to a large extent. During the last several years K. Dinker Rao and myself had worked together in several areas including the action research project Planning for Development of Villages (PDV) wherein we are experimenting with a new lending methodology. Our constant interactions have greatly influenced this work. Interactions with a large number of bankers also who have been associated with the above project has helped in formulating ideas. R. Bandyopadhyay, K.V. Patel, N.B. Shete, D.P. Khankhoje had offered their suggestions on several occasions. Y.S.R. Murthy had helped with few software packages. George Davis and Tommy Mathew had provided editorial help on several occasions. Vrinda Patharkar had helped with excellent research assistance, and carried out several exercises. Chandrakala Bhopalkar wordprocessed the entire text efficiently and Pushpa Talekar continued this job later. I am extremely thankful to all of them.

My thanks are also due to the staff members of National Institute of

Bank Management (NIBM) who have provided various kinds of assistance at various stages of this work. My sincere thanks go to the organization of NIBM which has provided all research facilities in last one decade. I would like to express my thanks to Bank of Baroda also for permitting me to include a revised version of the paper 'Alternative Approach to the Social Banking'.

Outside the professional circle, I had regular discussions on sociological dimensions and rural development with my wife Lopa. These informal discussions too have helped me immensely in the preparation and writing of this book.

None, however, can be held responsible for either any error which still remains or any unscientific view which has been expressed here. The author himself owns responsibility for all such lapses.

R. DAS GUPTA

CONTENTS

LIST OF TABLES

1

AGRICULTURAL PERFORMANCE

Introduction

As the whole purpose of agricultural credit is to improve the agricultural performance, it is important to analyse the latter first. It is also important to note that agricultural performance in a developing country like India strongly influences the total economy through both forward and backward linkages.

This performance is generally measured through growth rates. Debate on rate of agricultural growth in India which started in late sixties[1] is perhaps still important for analysis. Although the exercise earlier was to mainly evaluate the so called 'green revolution', in later period Vaidyanathan's[2] conclusion of deceleration in Indian agriculture was to explain the stagnation in the industrial growth in the Indian economy since mid-sixties. What is, however, interesting to note is that same set of data has yielded different results in different exercises depending upon the methodology adopted. All these have been very nicely summarised by Ray[3] who observes "some have shown through sophisticated measurement techniques, that new crop technology has helped to maintain, and in some cases, even accelerated output growth rates. Others have shown through equally complex measurement techniques that new technology has failed to accelerate, and in some cases has decelerated output growth rates".

Methodological Problems

If agricultural output had been close to any type of trend curve, there would not have been any room for controversy. Instead, agricultural output because of variations in weather to an unknown extent fluctuates tremendously which strictly speaking tends to follow some oscillating curves. We, however, generally take help of non-oscillatory type of functional forms to analyse the agricultural production data, and then estimate growth rate. One reason for this is the absence of enough 'friendly' packages to estimate such oscillating functions. Secondly, it is somehow difficult to talk about growth rates in that context as a single series will give both increasing and decreasing rate of growth. Problem however, can be circumvented by using not all data but few data like peak values, trough values or some other values. We will then have an idea of change in notional output. They, however, need not be notional outputs of normal weather.[4] One may in fact be interested in knowing the changes of output in worst weathers. We will then know the extent of our control over nature.

As already mentioned, there has been lot of controversy regarding the growth rate of agricultural production in pre- and post-green revolution periods in India. This has been mainly because of choice of different statistical functional forms or different cut off periods or different statistical specifications or combination of few or all of them. This is because, as Krishnaji[5] and Rudra[6] observe, (i) particular function although not truly representing the system may provide a good fit to a particular set of data, and (ii) everything turns on which hypothesis is treated as null hypothesis. Dasgupta[7] using (i) Dummy variable, and (ii) Piece-wise specifications and considering either 1965-66 or 1969-70 as cut off period observed that change in growth rate of agriculture in two periods depend upon specification and cutoff points.

Even if one ignores the problem of cutoff year, one may get contradictory result depending on model specification. Dasgupta[8] observed that only in about 30 per cent of cases (for the same cutoff year) results arising out of the two different specifications converged so that a conclusive inference could be drawn. In rest of the cases result of one model did not tally with that of other, and no inference could be drawn regarding change of growth rate.

Arithmetical Framework

It is, therefore, quite risky to use available statistical methodologies to infer about changes in growth rate. Krishnaji[9] is right to observe "these procedures have no theoretical base either economic or statistical, and consequently inferences on pattern of growth are not valid". He in fact, asks a valid question "why do we need a trend rate of growth? If actual output has fallen by 10 per cent over the preceding year, it is no comfort to be told that it would have risen by three per cent had normal weather prevailed". What one is actually interested is in knowing about performance of agriculture in different periods or a change in that between two periods. Issue of debate is whether agricultural sector had been growing satisfactorily or not. This can perhaps be explored through primitive arithmetical approach.

What we are therefore interested is in analysing change of agricultural performance between different periods in last thirty-six years from 1952 to 1987. Whole of 36 years has been divided into three subperiods of twelve years each, and then simple growth rate of annual averages of these periods has been calculated. This twelve year period is supposed to contain both good and bad years, and annual average production level of each period gives us an idea about notional output of that period or a decade. We can then compare the growth rate of notional outputs in two different time intervals. Secondly, we can afford to be less careful to choose a precise cutoff year. Whether deceleration comes into action in 1964-65 or in 1968-69, we would expect to get a reduced growth rate in the latter period. One can however, argue that this type of exercise will lead to meaningful comparisons of growth rates only if coefficients of variations of each period are of same order. This is true if we are measuring statistical growth rate in a very precise sense. If we are, however, interested in agricultural performance rather than statistical growth of agriculture over different periods, and considering the fact that (i) agricultural commodities can be stored as buffer stock (not without cost however), and (ii) and in the absence of large scale irrigation facility (or dependence on monsoon of random nature) farmer's investment will depend upon favourable weather, we think we will be able to judge the agricultural performance through the measurement of decennial growth rate. We are, however, since going to compare two average growth rates, we will not be able to discuss whether change in growth rate has taken place within a particular period. Even statistical models derived earlier don't capture this. We could have captured that through our method if we had more number of periods say of 5 years each. We will however, not do this since we are not interested in short run growth rates.

Secondly, if the span of periods is less we may again land up in comparing growth rates of production between 'good' and 'bad' periods. We are after all interested in the long-run tendency of agricultural production in India.

We have for discerning this long-term tendency used thirty-six years of production data from 1952-53 to 1987-88. These thirty-six years have been divided into three periods 1952-1963, 1964-1975,

1976-1987, of twelve years each. If we use the Yojna Report information (Table 1.1), both second and third periods seem to be equivalent in terms of weather. The first period on the contrary is a very good one with only one moderately bad year. Second period has six while third period has five bad years ranging from slightly bad to severely bad. Growth rate then has been estimated between two periods say second and first, and third and second. Average production of the period has been supposed to be concentrated in the middle of that period. This implies that length between two consecutive periods too is twelve years. The growth between first and third periods when divided by the length of years i.e. 24 gives us long-term simple annual growth. Whereas a low growth between first two periods may be attributed to bad phase of weather, the same between last two periods cannot be because of that. Instead of estimating growth rate of individual crops, growth rates of group of crops have been estimated. Growth rate for cereals, pulses, food grains and oil seeds have been estimated for physical level of production. These are unweighted growth rates. Index numbers using "All India" weights have been used to estimate the growth rate of agricultural output as well as cereals, pulses, foodgrains and oilseeds. Thirty-five crops considered in this exercise account for more than ninety-seven per cent of weights. Some of the states have been considered together. Punjab and Haryana; Maharashtra and Goa; Tamil Nadu and Pondicherry; and Assam and other North Eastern States have been aggregated to form four states only.

Growth Rates

Table 1.2 which gives growth rate (both weighted and unweighted) between first and third period gives an indication of long-run growth rate. If weighted growth rate is more than unweighted growth rate, it indicates that production shift has taken place in favour of high weighted or high valued crops within that group of commodities. For cereals, whereas production shift in favour of high weight crops has taken place in Andhra Pradesh, Punjab and Haryana, Rajasthan and Uttar Pradesh, opposite has happened in Bihar and Tamil Nadu. In rest of the states not much of shift seems to have taken place. For pulses, whereas in Andhra Pradesh shift has taken place in favour of high weight pulse crops, in Gujarat shift has taken place in favour of low weight pulse crops. For foodgrains as a whole, Andhra Pradesh, Maharashtra, Punjab and Haryana, Rajasthan and Uttar Pradesh have gone for production shift in favour of high weight crops; Assam, Orissa and Tamil Nadu for low weight crops. In case of oil seeds Bihar and Rajasthan have favoured high weight crops; and Gujarat, Madhya Pradesh, Punjab and Haryana and West Bengal have favoured low weight crops. This shift might have taken place because of either in shift of area or in relative change of yield *vis-a-vis* other crops.

TABLE 1.1 : Years of drought in India (1952-1987)

Year	Area afected × 10 sq km.*	Percentage of area of the country afffected*	Category**
1952	0.81	25.8	S
1965	1.35	42.9	SV
1966	1.01	32.3	M
1968	0.45	20.6	S
1969	0.62	19.9	S
1971	0.42	13.3	S
1972	1.39	44.4	SV
1974	0.92	29.3	M
1979	1.24	39.4	M
1982	1.04	33.1	M
1985	0.95	30.1	M
1986	0.60	19.0	S
1987	1.55	49.2	SV

* *Yojana*, June 16-30, 1989 Vol 33, No.11

** Percentage of area affected — Category
0-25 — S = Slight
25-40 — M = Moderate
40 and above — SV = Severe

If we now consider 4 per cent or above simple annual growth rate be high growth rate, 2-4 per cent moderate, and below 2 per cent a low level of growth rate when weighted growth rate is considered, Punjab and Haryana, Rajasthan and Uttar Pradesh have achieved high growth rates in foodgrains in last four decades (Table 1.3); Gujarat, Karnataka, Maharashtra, Punjab and Haryana, Rajasthan and Tamil Nadu have achieved high growth rates in total agricultural production (Table 1.4). All these high growth states, however, did not have similar performance in all three sub-periods. In case of food grains, whereas Punjab and Haryana achieved high growth rates for both the consecutive periods, Rajasthan and Uttar Pradesh achieved medium growth between first and second period, and high growth in last two periods. In case of total agricultural production only Punjab and Haryana and Gujarat achieved high growth rates for both the periods.

TABLE 1.2 : Long-run growth rates (1952-87)

(Per cent per annum)

Name of the State	Cereals		Pulses		Oilseeds		Foodgrains	
	W	UW	W	UW	W	UW	W	UW
Andhra Pradesh	3.8	2.5	2.7	2.2	1.9	1.6	3.8	2.5
Assam	2.9	3.3	–0.4	0.3	6.3	6.2	2.9	3.3
Bihar	1.9	2.7	–1.0	–0.9	3.6	2.1	1.9	2.0
Gujarat	3.9	8.5	5.0	6.3	2.3	3.1	3.9	3.7
Karnataka	3.1	3.2	2.2	2.2	0.8	0.9	3.1	3.1
Kerala	1.3	1.7	0.3	0.3	–1.2	–1.4	1.3	1.7
Maharashtra	3.0	2.7	0.9	0.7	–0.3	–0.2	2.9	2.4
Madhya Pradesh	1.8	1.9	1.8	1.7	0.4	0.9	1.8	1.9
Orissa	0.8	1.4	8.4	8.7	41.4	25.1	0.9	1.9
Punjab and Haryana	28.4	17.1	–2.5	–2.4	8.8	20.7	24.4	10.9
Rajasthan	5.5	3.3	1.5	1.9	10.3	8.1	5.3	2.9
Tamil Nadu	2.4	2.9	4.5	4.7	–0.04	–0.04	2.4	2.9
Uttar Pradesh	7.1	5.7	–1.3	–1.2	0.9	0.8	6.5	3.9
West Bengal	2.6	2.8	–1.3	–1.2	12.6	13.2	2.5	2.6

W = Weighted
UW = Unweighted

TABLE 1.3 : Simple annual growth rates of foodgrain production in last four decades

	Long-run (1952-87)	Phase I (1952-63/1964-75)	Phase II (1964-75/1976-87)
Anhra Pradesh	3.8	2.9	3.4
Assam	2.9	1.9	3.2
Bihar	1.9	1.8	1.6
Gujarat	3.9	3.1	3.6
Karnataka	3.1	4.5	1.1
Kerala	1.3	2.6	0.02
Maharashtra	2.9	0.6	4.9
Madhya Pradesh	1.8	0.7	2.6
Orissa	0.9	1.5	0.3
Punjab and Haryana	24.4	12.0	13.1
Rajasthan	5.3	2.9	5.8
Tamil Nadu	2.4	3.4	1.0
Uttar Pradesh	6.5	3.2	7.1
West Bengal	2.5	2.6	1.8

If we look into changes in growth rates or 'acceleration' and 'deceleration' we find that in case of foodgrains, acceleration has taken place in all the three high growth states; all but Karnataka, Tamil Nadu and West Bengal among medium growth states. And among low growth states, all except Madhya Pradesh face deceleration. In case of total agricultural production, all high growth states except Karnataka and Tamil Nadu achieved acceleration. Among medium growth states only Assam and Uttar Pradesh have achieved acceleration. And in the last category, only Madhya Pradesh has achieved acceleration.

TABLE 1.4: Simple annual growth rates of agricultural production in last four decades

	Long-run (1952-87)	Phase I (1952-63/1964-75)	Phase II (1964-75/1976-87)
Andhra Pradesh	3.5	3.4	2.5
Assam	3.1	2.1	3.5
Bihar	1.1	1.1	0.9
Gujarat	6.6	4.2	5.9
Karnataka	5.0	5.5	2.7
Kerala	1.1	2.4	–0.1
Maharashtra	5.1	2.2	6.3
Madhya Pradesh	1.7	0.8	2.4
Orissa	1.4	1.9	0.6
Punjab and Haryana	13.0	8.3	8.9
Rajasthan	5.6	3.5	5.6
Tamil Nadu	4.5	5.1	2.4
Uttar Pradesh	3.6	1.8	4.5
West Bengal	2.5	2.5	1.9

TABLE 1.5 . Relationship between long-run growth and change in growth (foodgrains production)

Long-run growth	Change in growth	
	Acceleration	Deceleration
High (>4)	Punjab and Haryana Rajasthan Uttar Pradesh	—
Medium (2-4)	Andhra Pradesh Assam Gujarat Maharashtra	Karnataka Tamil Nadu West Bengal
Low (<2)	Madhya Pradesh	Bihar Kerala Orissa

What therefore, emerges from above observations is that only a few states have shown satisfactory growth performance in agricultural production. These are Punjab and Haryana, Gujarat, Rajasthan, Maharashtra, and also Uttar Pradesh. Karnataka and Tamil Nadu, although high growth states are facing deceleration. Among other states Assam and Madhya Pradesh seem to be performing satisfactorily although they are medium and low growth states. As it appears that acceleration is more pronounced among high growth states, deceleration cannot be considered as a result of higher production level.

TABLE 1.6 : Relationship between long-run growth, and change in growth (agricultural production)

Long-run growth	Change in growth	
	Acceleration	Deceleration
High (>4)	Punjab and Haryana Rajasthan Maharashtra Gujarat	Karnataka Tamil Nadu
Medium (2-4)	Assam Uttar Pradesh	Andhra Pradesh West Bengal
Low (<2)	Madhya Pradesh	Bihar Kerala Orissa

Fluctuations

Another aspect, apart from growth, for evaluating the agricultural performance is stability in production. This is more true in our case, as we have ignored distributional aspect while estimating growth between two periods. For measuring the stability we have estimated fluctuation error which has been given below :

$$\text{Fluctuation Error} = \sum_{t=1}^{12} \frac{|Yt - E(Yt)|}{|E(Yt)|}$$

where *Yt* is production level in *t*th year, and *E(Yt)* is expected or trend level of production in *t*th year.

In $Yt = a + bt$ has been fitted as a regression trend line. Compound growth rates and fluctuation errors have been presented in Tables 1.7 and 1.8. One can note that so far as long-run growth rates are concerned, compound growth rates in Table 1.7 and simple growth rates (Tables 1.3 and 1.4) computed through arithmetical procedure reveal similar pattern.

TABLE 1.7 : Long-run compound growth rates (1952-87)

Name of the State	Foodgrains	Agriculture
Andhra Pradesh	2.7	2.6
Assam	2.2	2.4
Bihar	1.7	1.1
Gujarat	2.5	3.9
Karnataka	2.4	3.5
Kerala	1.2	1.1
Maharashtra	2.2	3.3
Madhya Pradesh	1.6	1.6
Orissa	0.9	1.2
Punjab and Haryana	8.2	6.0
Rajasthan	3.3	3.4
Tamil Nadu	2.0	3.2
Uttar Pradesh	3.9	2.6
West Bengal	1.9	2.0

TABLE 1.8 : Measure of instability (fluctuation error)

Name of the State	Foodgrains	Agriculture
Andhra Pradesh	0.37	0.35
Assam	0.36	0.34
Bihar	0.67	0.53
Gujarat	1.48	0.95
Karnataka	0.59	0.45
Kerala	0.50	0.46
Maharashtra	0.71	0.57
Madhya Pradesh	0.64	0.62
Orissa	0.54	0.49
Punjab and Haryana	0.49	0.28
Rajasthan	0.85	0.86
Tamil Nadu	0.53	0.44
Uttar Pradesh	0.47	0.38
West Bengal	0.43	0.43

So far as instability is concerned, it is little more in case of foodgrains production compared to total agricultural production. Whereas Punjab and Haryana, and Assam have low instability in case of total agricultural production, no state can be classified under this category in case of foodgrains production. Higher instability, however, does not seem to be related with high growth rate (Tables 1.9 and 1.10). States like Punjab and Haryana, Karnataka, Maharashtra and Tamil Nadu while achieving higher growth rates have maintained low or medium instability. On the other hand low growth states like Bihar, Kerala, Orissa and Madhya Pradesh have experienced medium to high amount of instability.

TABLE 1.9 : Relationship between growth and variability (foodgrains)

	Fluctuation Error (Per cent)		
Compound Growth (Per cent)	>60 (High)	35-60 (Medium)	<35 (low)
> 3 (High)	Rajasthan	Punjab and Haryana Uttar Pradesh	—
1.75 - 3 (Medium)	Gujarat Maharashtra	Andhra Pradesh Assam Karnataka Tamil Nadu West Bengal	—
< 1.75 (low)	Bihar Madhya Pradesh	Kerala Orissa	—

TABLE 1.10 : Relationship between growth and variability (Agriculture)

	Fluctuation Error (Per cent)		
Compound Growth (Per cent)	>60 (High)	35-60 (Medium)	<35 (low)
> 3 (High)	Gujarat Rajasthan	Karnataka Maharashtra Tamil Nadu	Punjab & Haryana
1.75 - 3 (Medium)	—	Andhra Pradesh Uttar Pradesh West Bengal	Assam
< 1.75 (low)	Madhya Pradesh Kerala Orissa	Bihar	—

Performance

At the end, we would like to utilise three indicators long-run growth rates, change in growth rates and fluctuation error to evaluate the agricultural performance of different Indian states in last four decades.

We have three levels of long-run growth High (H), Medium (M) and Low (L); two types of change in growth rates between different periods - Acceleration (A) and Deceleration (D); and three levels of instability High (H_1), Medium (M_1) and Low (L_1). Based on these indicators, agricultural performance has been classified under five categories which have been presented in Table 1.11. Each state then has been classified according to its performance in Table 1.12.

TABLE 1.11 : Classification of agricultural performance

Level	Combination of indicators
Excellent	HAL_1, HAM_1
Good	HAH_1, MAL_1, MAM_1
Average	HDL_1, HDM_1, HDH_1, MAH_1, MDL_1
Poor	MDM_1, MDH_1, LAL_1 LAM_1, LAH_1, LDL_1
Very Poor	LDM_1, LDH_1

Note: HAL_1 indicates High (*H*) level of long-run growth, Acceleration (*A*) and Low instability (L_1)
LDH_1 indicates Low (*L*) level of long-run growth, Deceleration (D) and High instability (H_1)

What we thus observe is that eight states have similar level of performance for both foodgrains and agricultural production. Whereas Punjab and Haryana has done excellently well in both foodgrains and agricultural production, Bihar, Kerala and Orissa have shown very poor results in case of both foodgrains and agriculture. Among other states Maharashtra has although been excellent in total agricultural production, it has shown average result in foodgrain production. On the other hand Uttar Pradesh has shown excellent result in foodgrains production, but has been just good in total agricultural production.

TABLE 1.12 : Agricultural performance

	Foodgrains	Agriculture
Excellent	Punjab and Haryana Uttar Pradesh	Maharashtra Punjab and Haryana
Good	Andhra Pradesh Assam Rajasthan	Assam Gujarat Rajasthan Uttar Pradesh
Average	Gujarat Maharashtra	Karnataka Tamil Nadu
Poor	Karnataka Madhya Pradesh Tamil Nadu West Bengal	Andhra Pradesh Madhya Pradesh West Bengal
Very Poor	Bihar Kerala Orissa	Bihar Kerala Orissa

Notes and References

1. Ashok Mitra (1968)
2. A. Vaidyanathan (1977)
3. S.K. Ray (1991)
4. N. Krishnaji (1980)
5. *Ibid.*
6. A. Rudra (1982)
7. R. Dasgupta (1986)
8. *Ibid.*
9. N. Krishnaji, *Op. cit.*

2

DETERMINANTS OF AGRICULTURAL PERFORMANCE

Introduction

Agricultural output is basically a technological function of inputs and natural environment – a totality of temperature, wind, soil fertility, humidity, etc. If this environment be considered as given, or not subject to change in short-run, agricultural output or yield rate depends upon application of input and technology used. Application of these forces of production, however, depends upon economic environment which in Indian case is a combination of overall domestic market condition and production relation in agricultural sector. These are sufficient conditions of agricultural performance where as technological function is just a necessary condition.

Framework

Production relation is a crucial determinant. In Bhaduri's[1] terminology semifeudalism in most of the eastern part is the reason of agricultural backwardness of that area. According to that framework a big landlord simultaneously operates in four markets, land by leasing it out, labour by hiring labour for agricultural operations, money by providing credit to tenants and agricultural labourers, and trading by buying agricultural products within the village and selling that inside and outside the village. The big landlord then allocates his fund in all these markets, and the proportion is determined by the short-run return. Agricultural activity being the least rumunerative gets least attention. Consequently if there is less investment in irrigation, fertiliser or HYV seeds, agricultural growth rate remains slow. Following Patnaik[2], above agrument can be formalised thus :

Let R = Rent from land
P = Return from usury
A = Profit rate in direct capitalistic investment in agriculture
T = Return from trade
M = Total amount of money invested

Then the landlord will go for capitalistic investment in agriculture only if–

$$A > (R+P+T)/M \tag{1}$$

To understand the depressing[3] effect of above markets, let the model be formulated more explicitly using Basu's[4] framework.

A tenant produces X units of agricultural goods which depend upon technological factors, and keeps with him a proportion 'a' given by custom and determined by historical facts. Both are constants in the short-run. Bt be the amount borrowed by the peasant in year t and let Ct be the amount consumed by him. The interest rate is i. If the peasant does not save, then

$$Bt = Ct - \{ax - (1+i)B_{t-1}\} \tag{2}$$

In a stationary state when $Bt = B$, and $Ct = C$

$$B = C - \{ax - (1+i)B\}$$
$$\text{or } B = (a/i)x - (c/i) \tag{3}$$

Let $x = 100$ units (production), $a=1/2$ (50:50 crop sharing), and $i=1$ (100 per cent rate of interest) and $C = 30$ units (consumption).

Then B equals to 20 units. In that situation, in the following year when production is again 100 units (assumed to be constant), 50 units will be paid to landlord as rent, 40 units as repayment (principal plus interest), and the tenant will be left with 10 units. His consumption needs being 30 units, he will have to borrow again 20 units, thus pushing himself to a perpetual debt trap.

If the tenant reduces his consumption for some years, he can reduce the borrowing, and finally can come out of the trap. Alternatively, if the production be raised, his burden on borrowing is reduced (assuming C to be constant), and gradually it becomes nil. But neither of them seems to be possible. Since consumption *(C=30)* is already at subsistence level, it may be difficult to go below this mark. Even if they dare to suffer in some periods, it will be a futile attempt which we will see later.

Production cannot be raised as tenants are not in a position to invest. Long-term investment in irrigation or land development is anyway a risky proposal because of lack of security. Even if he borrows money from informal market (institutional credit may not be accessible), the investment may not be profitable because of high rate of interest. On the other hand whereas the landlord inspite of having enough fund of his own, and also easy access to formal credit market with low interest rate does not go for investment for two considerations – short-run and long-run.

Equation (1) depicts a short-run or a static phenomenon whereby return from agriculture is not very lucrative compared to those from other channels. Equation (2) tells the long-run or a dynamic story. If the landlord invests in agriculture, and production increases as a result of which income of the tenant also increases, tenant's need of consumption loan will decline, and eventually it will reduce to nil unless the tenant increases his consumption level. This will in the long-run deprive the landlord of a crucial source of income. It is not a question of economic gains only. Rural rich uses their large power, key to which is the dependence of the poor on the rich, to strengthen their position further, and therefore are not likely to encourage any policy measure that improves the conditions of the poor.[5] There is therefore, no technological innovation, and stagnation in agricultural sector.

Above analysis has been debated on both theoretical and empirical grounds. Bardhan[6] does not find his large scale survey data of eastern India to be consistent with above hypotheis. He finds in his survey of four states viz., West Bengal, Uttar Pradesh, Bihar and Orissa that about 50 per cent of tenants seek production loan from landlords, and almost half of this is transacted without any interest charge. Not only that, tenants felt there that money-lending was not the principal occupation of their landlords. Nobody, however, denies the fact that money-lending is not the principal occupation of a landlord. It is rather a means to both increase the return on their investment level, and to keep their power over smaller people undisturbed. Even if it is granted that half of the loan were interest free (which need not be true if lower price of grains during harvest time is considered), this so called "goodwill" may be used other times (say election or policy issues) when decisions generally go against the interest of tenants, smaller farmers and other poorer lots of the village who are very much dependent on big landlords. Vaidyanathan[7] takes another view when he argues that Punjab and Haryana inspite of having high tenancy recorded high growth rate, and eastern India in early 1970s had only 25 per cent of operated area under tenancy. So far as empirical question is concerned it is difficult to rely on these estimates as Hossain[8] writes" The Planning Commission's task force on Agrarian Relations in 1973 found that only 0.8 million share cropping agreements out of 2.2 million were legally recorded in the state of West Bengal". This is corroborated by Bardhan's observation too that landlord often rotates his tenants so that they do not acquire occupancy right in the land and shift the same worker from tenancy to wage labour and back to tenancy.[9] So far as evidence of usury is concerned Prasad's[10] data from Bihar in 1981-82 shows that about 93 per cent of agricultural labourers do take loan from their employer, and only 10 per cent of cases are without interest rate. Otherwise, in more than 75 per cent of cases interest charge is more than 36 per cent, infact about 40 per cent labourers paying more than 60 per cent of rate of interest. Again, collecting data from West Bengal in 1976-77 Bandyopadhyay[11] concludes that landlord is more interested in exploiting tenants by imposing higher rates of interest on loans to them than in undertaking direct productive investment. In fact Bardhan's[12] own data elsewhere shows that most of the consumption and general purpose loan provided by landlords in eastern India carry interest rate. Bardhan[13], however, may argue that inspite of usury, more than 60 per cent of his landlords with tenants use chemical fertilisers and HYV seeds, thus taking part in productive investment. It will be perhaps too naive to argue that landlords engaged in money-lending to his tenants and labourers will not use chemical fertilisers or HYV seeds at all. It would have been a very conclusive evidence in favour of his argument if he had shown that level of use of chemical fertiliser or HYV seed by these landlords is more or at least same than that by cultivators of similar size without money-lending activity. What finally seems is that activity of usury prevails in Indian, atleast eastern Indian agricultural economy, and Bhaduri's model cannot be invalidated on empirical ground.

Earlier model has however, been contested on theoretical ground too. Newbery[14] argues that usury is a very weak constraint on innovation, and accepting the exploitative framework of Bhaduri he puts forward that landlords have sufficient power to extract the extra gain from innovation by suitably manipulating the rental share, the interest rates etc. Ghose and Saith[15] in fact go further by developing a model whereby stronger the domination of the landlord, the more readily he accepts technological improvement.

There is definitely a substantial merit in above arguments. Because of this domination tenants don't have incentive for either reducing consumption or increasing production (equation 2) to come out of the debt trap. Tenants know about landlord's power too well, and find it a futile attempt to get rid of it. They will always be brought back to the trap. This however, does not justify Ghosh and Saith's contention that because of stronger domination landlords will go for technological improvement and agricultural growth. They are in fact constrained by profitability and market.

This is the central point of our argument, and we agree with Bardhan's[16] criticism that tenancy is uncritically equated with feudalism. We would like to consider capitalism and feudalism more as psychological behaviour. Ignoring all theoretical compulsions we would like to equate capitalism with Schumpetarian entrepreneurship whereby a capitalist is mainly concerned with long-term considerations, and hence makes innovations and investments. He is not like the commercial class who largely aims at controlling the organisation of marketing, but not of production. He may either engage himself in self cultivation, or have tenants, or manage his farms by both the means depending upon his manpower, managerial capability etc. His production decision is not governed by the short-run market or profit rate but by long-run market and total profit. He may like to find out a new crop and a new market. New market may already exist in international segments, or may be created within the economy by helping in employment generation through proper technological choice in agriculture, or by increasing the purchasing power of rural population through unexploitative production relations. In Japan, Taiwan and most parts of mainland China far sighted landlords helped the tenants through advances of working capital, and initiatives for improving the infrastructure and related farm practices.[17]

Other kind of farmers on the contrary rely on short-run market. Having a myopic view, exploitative production relation may be considered by them as one of the most important tools for profit making. This is what we would like to call feudalistic behaviour. This is immaterial whether they go for self cultivation or engage tenants. In case of self cultivation agricultural labourer will be exploited, and in latter case tenants will be exploited. Tenancy *per se* is therefore not a necessary condition of exploitation or agricultural stagnation.

Because of this narrow market higher agricultural growth does not take place, and in the absence of higher agricultural growth domestic market does not expand (ignoring the technological problem which itself may be labour saving)

We have thus

Narrow home market-----> low agricultural growth-------> smaller home market------> low economic growth--------> low agricultural growth.

This market trap, however, can be tackled by some exogenous factors viz. government or credit agencies, taking care of poorer people with both production and consumption loan to free them from exploitative informal money market and increasing their purchasing power or by industrial growth. Effect of industrial growth on agricultural growth will be more prominent if industrial expansion is more employment generating. This not only creates a market in short-run sense but creates a condition whereby exploitative relations in rural areas weaken. One of the positive aspects of Employment Guarantee Scheme (EGS) of Maharashtra has been that it compelled the private farm owners to offer higher wage rates.[18]

It is therefore argued that technological factors, psychological conditions, role of government and credit institutions and size of market are determinants of agricultural performance.

Role of government in this context has been perceived through canal irrigation system : its growth and effective utilisation of water. Canal irrigation not only provides water directly to crops, but raises water table of the area and helps in the growth of well irrigatin system. Other important roles, however, are providing infrastructure, extension service, effective intervention in the market process, preventing exploitation etc.

Credit through formal lending institutions can play tremendous role in improving the agricultural performance two ways. It directly helps in increasing inputs of fertiliser, pesticides, HYV seeds etc., and irrigating more areas through wells. It can also ease the burden of farmers indebted to money-lenders. Formal credit then indirectly helps to raise the real income of indebted farmers, and more investment may be done in the agricultural sector by these farmers for further growth.

Also the money-lenders, specially landlord money-lenders, now being deprived of usury income may be forced to invest in productive sector to compensate for the loss of income.

We therefore intend to

(i) examine the variations in output in each state explained by variations in different inputs;
(ii) study the impact of variables like soil fertility, rainfall, fertiliser use, HYV area, market size etc.on agricultural production; and
(iii) evaluate the influence of different factors specially institutional credit on fertiliser use, HYV area and well irrigation.

Data Base

We have collected both cross-sectional and time series data for above analysis. Data requirement for such an exercise are on (i) agricultural production, (ii) area, (iii) use of input, (iv) rainfall, (v) market size, and (vi) disbursement of credit.

We have production data of 35 major crops for 36 years from 1952-53 to 1987-88 for seventeen states; irrigation statistics from 1965-66 to 1982-83; land utilisation statistics from 1966-67 to 1982-83; fertiliser data from 1967 to 1984; HYV statistics from 1966-67 to 1984-85; annual rainfall data for 37 stations from 1966 to 1984.

For rural credit we collected data pertaining to three agencies viz. commercial banks, co-operative societies, and land development banks. We required disbursement data for our analysis. This data for commercial bank was however available only from 1978 till 1984. Only outstanding data are available for the period 1969 to 1977. In recent years there has not been any stable relationship between outstanding and disbursement. In earlier years i.e. 1977 and 1978, there was, however, a stable relationship between outstanding and disbursement for each state. We therefore, used this co-efficient (given in Appendix-I) to estimate the disbursement of agricultural credit by commercial banks between 1969 to 1977. It also should be noted that weightage of commercial bank in total agricultural credit was quite small in earlier years.

Rural credit data (disbursement) by Primary Agricultural Credit Societies (PACS) and Land Development Bank (LDB) has been available from 1963-64 to 1982-83. There has been problem of compatability between credit data of commercial banks on one hand and that of PACS and LDB on the other hand. Former gives April to March data whereas latter gives July to June data. We have, however, ignored these discrepancies and added yearly disbursement data by each agencies to find out total agricultural credit in a year. Purposewise agricultural credit too has been estimated similarly.

All these different types of data, for different time spans pertain to different geographical coverages. We have therefore, aggregated different data to have statistics for 14 states viz. (1) Andhra Pradesh (AP), (2) Assam (including Nagaland, Mizoram, Tripura, Arunachal, Meghalaya) (3) Bihar (4) Gujarat (5) Karnataka (6) Kerala (7) Maharashtra (including Goa) (8) Madhya Pradesh (MP) (9) Orissa (10) Punjab (including Haryana) (11) Rajasthan (12) Tamil Nadu (including Pondicherry) (13) Uttar Pradesh (UP) and (14) West Bengal(WB). Because of data constraint we have confined our analysis to post green revolution period from 1966-67 to 1982-83.

Production data have been aggregated to form indices. Volume index of foodgrain has been formed by simply adding the physical production of foodgrain crops produced in a single year. Weighted agricultural index has been calculated considering all the crops and all India weights assigned to each crop (Weights being given in Appendix-II).

Technological Variation

We first attempted to estimate the technological variation in the agricultural production because of different inputs. Independent variables considered have been fertiliser level, irrigation, HYV area and annual rainfall. Regression equations have been estimated with time series data. To overcome the problem of auto-correlation maximum likelihood estimation method has been used assuming an autoregressive series of order one.

Following four functions in double log form have been estimated.

(i) $VA = f(V2, V3, V4, V5)$
(ii) $VA = f(V2, V6, V4, V5)$
(iii) $VF = f(V2, V3, V4, V5)$
(iv) $VF = f(V2, V6, V4, V5)$

where VA = Weighted Agricultural Index

VF = Volume index of food
*V*2 = Fertiliser (NPK) per unit area
*V*3 = Total Irrigated Area
*V*4 = Proportion of HYV area
*V*5 = Annual rainfall
*V*6 = Well irrigated area

Multiple correlation co-efficient (R^{-2}) for all four types of function for each state have been given in Table 2.1. What is evident is that amount of variations explained by the four independent variables differ quite significantly from state to state. One of the important variables excluded from the above specification is climatic factor which however should not have changed significantly in the span of these 17 years. Pesticide use too has not been considered. Annual rainfall may not be a very appropriate variable. Timely rainfall should have been a better variable. Effect of flood, and parasite should also have been considered. Nevertheless it seems that these variations reflect a kind of efficiency parameter prevailing in different regions. These functions were not estimated for West Bengal as reliable irrigation data were not available. Whereas more than 75 per cent of variation in agricultural production is explained by these four variables in Andhra Pradesh, Gujarat, Karnataka, Maharashtra, Madhya Pradesh, Punjab and Haryana, Rajasthan and Uttar Pradesh, less than 50 per cent of variation is explained by these variables for Bihar, Kerala, and Orissa. Also these latter states are ones with low growth and poor performance.

TABLE 2.1: Technological variation (R^2)

Functional Form / State	1	2	3	4
1. Andhra Pradesh	.85	.85	.83	.83
2. Assam	.68	.63	.67	.59
3. Bihar	.46	.46	.42	.42
4. Gujarat	.79	.79	.72	.72
5. Karnataka	.85	.83	.67	.60
6. Kerala	.37	.36	.46	.46
7. Maharashtra	.96	.96	.90	.91
8. Madhya Pradesh	.86	.87	.81	.81
9. Orissa	.41	.42	.56	.55
10. Punjab and Haryana	.94	.94	.93	.93
11. Rajasthan	.83	.81	.53	.55
12. Tamil Nadu	.46	.61	.59	.71
13. Uttar Pradesh	.85	.85	.92	.91
14. West Bengal	—	—	—	—

Either instability in production because of nature of crop, or environmental factor is dominant in these regions or these regions lack in efficiency in input output relationship which is a reflection of ethos of those regions. In Rajasthan whereas agricultural production is explained well by the technological function, food production is not, which may be because of instability in production because of vagaries of weather. Karnataka too shows a similar picture.

Analysis of Output

Cross-Section

Cross-sectional analysis has been carried out to find out the influence of endogenous and exogenous variables on agricultural performance. Two exogenous variables are natural soil fertility and rainfall (which, it may be argued, after all is dependent upon environment which is subject to change because of man, and so an endogenous variable). They have been considered exogenous in the sense that they cannot be easily influenced by man in a short-run. Another independent variable, subject to man's decisions is irrigation which induces for fertiliser and HYV inputs. These variables are, however, endogenous as investment decisions are taken considering so many factors viz., profitability, market size, availability of credit etc. Well irrigation is generally taken as substitute for canal irrigation. Experience in Maharashtra, however, shows that unusual growth of well irrigation has taken place in command areas.[19] Well irrigation therefore even in all India situation is expected to be more where proportion of irrigation (through canal) is also more. And finally these investment decisions are more positive where purchasing power is more because of larger market.

We therefore propose to estimate following linear regression functions:

(v) $V1 = f(V2, V5,)$
(vi) $V1 = f(V2, V4)$
(vii) $V1 = f(V6, V4)$
(viii) $V1 = f(V3, V4)$
(ix) $V1 = f(V2, V4, V7)$
(x) $V1 = f(V2, V5, V7)$

where

*V*1 = Agricultural Index/Net sown Area (NSA)
*V*2 = Total irrigation/NSA
*V*3 = Total Fertiliser (NPK) input/NSA
*V*4 = Annual Rainfall
*V*5 = Total Productive Workers/NSA (Effective Market)
*V*6 = HYV (cereals) Area/NSA
*V*7 = Soil fertility

This cross-sectional analysis has not been carried out for a particular year. We have instead, divided the whole span of 18 years into three phases consisting of equal number of years i.e. 6 years. Simple average of each variable has then been estimated for different phases for all 14 states.

In the absence of any measure of soil fertility, we have taken average yield rate of cereals of three years 1952-55 as an index of soil fertility. Instead of all agricultural crops, only cereals crops have been considered because of problem of homogeneity for aggregation. As the yield during that period was very much dependent on natural factors (in the absence of HYV, fertiliser, pesticides, irrigation) etc. this yield rate has been considered as an indicator of soil fertility.

Instead of considering all work force for our "market" we have estimated effective market as

Effective Market = Total Work force — Agricultural labourers.

Because of problems of underemployment, agricultural labourer is a kind of reserve army of labour, and accepts subsistence wages, and may not offer an effective market. This however, does not suggest that the whole class of agricultural labourer is irrelevant for any effective agricultural market. But to understand the variation in agricultural performances, it is better if only that segment of workforce be considered which constitutes a market (through purchasing power) with certainty. These figures of total workforce and agricultural labourers have been taken from census statistics of 1961, 1971 and 1981. 1971 and 1981 figures have been considered for our Phase 2 and Phase 3 data base. For Phase 1, average of 1961 and 1971 figures have been considered.

Regression results shows that agricultural productivity depends on soil fertility and rainfall. It has generally been found that productivity is more in high rainfall areas. It has, however, little weaker impact in phase 3 i.e. in recent years. Soil fertility is still an important variable to explain the agricultural productivity inspite of the fact that dominance of fertiliser, irrigation and HYV has increased. These three variables too have been significant determinant in all three phases. Fertiliser input, however, had not been a strong factor in phase 1. What R^{-2} suggests is that importance of HYV and fertiliser has been more in Phase 2 and Phase 3. Apart from these technological variables, the other economic variable which has been important in explaining the agricultural productivity is the effective market. This variable has been included in equations (1) and (6). Where as equation (1) suggests strong correlationship of this variable with agricultural productivity for all three phases, equation (6) suggests strong relationship in first two phases only. If this weakness in equation (6) contrary to that in equation (1) is not because of any statistical problem (multicollinearity), it may then be due to less dependence on local market in recent years because of improved communication system and production related to not local market but national market. According to equation (1) local market, however, is a strong determinant of agricultural productivity in Phase 3 also. We could have much more stronger proof to our hypothesis if we could correlate the agricultural growth with employment growth (or growth of market) which we could not do because of lack of data.

TABLE 2.2a: Multiple Regression Result for Agricultural Productivity (Regression Co-efficients and R^{-2})

(PHASE 1)

Variables	Equations					
	1	2	3	4	5	6
1. Proportion of irrigated area(V2)	44.9[a]	70.4[a]	—	—	37.2[a]	20.7
2. Fertiliser input (NPK) per unit area (*V3*)	—	—	—	455.3	—	—
3. Proportion of HYV area (*V6*)	—	—	49.9[a]	—	—	—
4. Annual Rainfall (*V4*)	—	.01[a]	.008	.007	.007[a]	—
5. Soil fertility (*V7*)	—	—	—	—	9.9[a]	9.6[a]
6. Effective Market (*V5*)	11312.2[a]	—	—	—	—	7410.8[a]
R^{-2}	.79	.79	.40	.28	.92	.91

a = significant at 5% level

TABLE 2.2b: Multiple Regression Result for Agricultural Productivity (Regression Co-efficients and R^{-2})

(PHASE 2)

Variables	Equations 1	2	3	4	5	6
1. Proportion of irrigated area (*V2*)	51.9[a]	72.0[a]	–	–	37.3[a]	28.2[a]
2. Fertiliser input (NPK) per unit area (*V3*)	–	–	–	559.3[a]	–	–
3. Proportion of HYV area (*V6*)	–	–	137.9[a]	–	–	–
4. Annual Rainfall (*V4*)	–	.008[a]	.009[a]	.007	.004[a]	–
5. Soil fertility (*V7*)	–	–	–	–	13.9[a]	12.2[a]
6. Effective Market (*V5*)	14945.9[a]	–	–	–	–	9890.5[a]
R^{-2}	.79	.65	.73	.47	.87	.91

a = significant at 5% level

TABLE 2.2c: Multiple Regression Result for Agricultural Productivity, (Regression Co-efficients and R^{-2})

(PHASE 3)

Variables	Equations 1	2	3	4	5	6
1. Proportion of irrigated area (*V2*)	66.6[a]	85.4[a]	–	–	55.8[a]	47.0[a]
2. Fertiliser input (NPK)per unit area (*V3*)	–	–	–	345.5[a]	–	–
3. Proportion of HYV area (*V6*)	–	–	127.3[a]	–	–	–
4. Annual Rainfall (*V4*)	–	.008[a]	.008[a]	1.485	.005	–
5. Soil fertility (*V8*)	–	–	–	–	11.6[a]	11.2[a]
6. Effective Market (*V5*)	9920.0[a]	–	–	–	–	4312.1
R^{-2}	.77	.76	.74	.58	.88	.87

a = significant at 5% level

Time Series

We now move to estimate effects of canal irrigation, well irrigation, fertiliser, and HYV area, on production index, either agricultural or foodgrains with time series data. Tables 2.3 to 2.6 present multiple regression results with OLS estimation procedure. Fertiliser and HYV inputs are most important variables to explain the production of both foodgrains and total agriculture. Influence of rainfall is mostly visible in Gujarat, Madhya Pradesh, Orissa and Rajasthan. Proportion of canal irrigation, however, has not been significant in any equation. Considering correlation coefficient (r) of 0.7 between two explanatory variables (so that r^2=0.5) multicollenearity seems to exist in about half of the states. Problem of autocorrelation too seems to exist in half of the states. It is, therefore, prudent to estimate single variable equations.

TABLE 2.3: Multiple Regression Results (OLS) (Dependent Variable : Agricultural Production Index)

Name of the State	Coefficients			Multi Collinear -ity	Auto Corre- lation	R^{-2}
	V(1)	*V*(2)	*V*(3)			
Andhra Pradesh	25344*	–1726	.02	N	I	.83
Assam	38208*	–1827	.08	N	I	.49
Bihar	11738	–2989	.79	N	N	.39
Gujarat	13255*	.702	.39*	N	N	.81
Karnataka	17508*	–1816	.19	Y	I	.79
Kerala	170	46	.02	N	A	–.05
Maharashtra	72374	1759	.58	N	I	.76
Madhya Pradesh	52782*	1539	1.49*	N	N	.85
Orissa	23904	–62	.57*	Y	N	.42
Punjab and Haryana	14953*	–2605	1.02*	Y	I	.91
Rajasthan	29379*	1013	.58*	N	I	.83
Tamil Nadu	12413*	–5766	.65	N	N	.44
Uttar Pradesh	48009*	–33155	1.50	Y	I	.85
West Bengal	—	—	—	—	—	—

Note :
V(1) : Total fertiliser per unit of Net Sown Area
V(2) : Proportion of Canal Irrigation Area
V(3) : Actual Rainfall
Multicollinearity N No
Y Yes
Auto Correlation N No auto correlation
A Positive significant auto correlation
–A Negative significant auto correlation
I Inconclusive
* Significant at 5 % level.

Single variable regression results with OLS estimation procedure have been given in Tables 2.7 and 2.8. All three important inputs viz., irrigation, fertilizer and HYV seeds have been significant variables in almost all states. Problems of auto-correlation, however, exists in about 40 per cent of cases. For these cases Cochrane-Orcutt method assuming first order autocorrelation has been used, and results have been given in Tables 2.9 and 2.10.

What we now find is that HYV is a significant determinant in all states except Kerala for foodgrain; all states except Kerala and Tamil Nadu for agricultural production. Fertiliser is significant in all states except Bihar, Kerala, and West Bengal for foodgrains; all states except Kerala and Rajasthan for agriculture. Well irrigation is significant in all states except Assam and Kerala for foodgrains; all states except Bihar and Kerala for agriculture. Canal irrigation is significant in all states except Assam, Kerala and Tamil Nadu for agricultural production. There are, however, few estimation problems for which we get negative results for few states.[20]

TABLE 2.4 : Multiple Regression Results (OLS) (Dependent Vatiable : Agricultural Production Index)

Name of the State	Coefficients			Multi Collinear -ity	Auto Corre- lation	R^{-2}
	V(2)	V(3)	V(4)			
Andhra Pradesh	–1691	.37*	.25	N	I	.81
Assam	195	.38*	–.02	Y	I	.65
Bihar	–2197	.10	.79*	N	N	.44
Gujarat	2293	.19*	.41*	Y	I	.72
Karnataka	–1707	.38*	–.02	N	N	.82
Kerala	1.072	.06	.01	N	A	.02
Maharashtra	2785 *	.39*	.27	Y	I	.92
Madhya Pradesh	4482	.22*	1.28*	Y	I	.85
Orissa	–93	.24*	.59*	Y	N	.52
Punjab and Haryana	–1006	.35*	.62	N	I	.86
Rajasthan	674	.12*	.62*	N	N	.77
Tamil Nadu	–12006	.01	.30	Y	I	.09
Uttar Pradesh	–23870	.41*	1.73	N	I	.83
West Bengal	—	—	—	—	—	—

Note :
- V(2) : Proportion of Canal Irrigation
- V(3) : Actual Rainfall
- V(4) : Proportion of HYV area

Multicollinearity	N	No
	Y	Yes
Auto Correlation	N	No auto correlation
	A	Positive significant auto correlation
	–A	Negative significant auto correlation
	I	Inconclusive

* Significant at 5 % level.

What we therefore, observe that all these three input variables have contributed to agricultural production in almost all states except Kerala. Only the extent of effect, and efficiency factor may be different. As a result some of the states inspite of having high growth in inputs have not experienced commensurate growth in production.[21]

TABLE 2.5 : Multiple Regression Results (OLS)
(Dependent Variable: Foodgrains Volume Index)

Name of the State	Coefficients			Multi Collinear -ity	Auto Corre- lation	R^{-2}
	V(1)	V(2)	V(3)			
Andhra Pradesh	1400 *	–145	.02	N	N	.76
Assam	5799 *	–247	.01	N	I	.47
Bihar	920	138.9	.04	N	I	.34
Gujarat	2251 *	367	.15*	N	I	.74
Karnataka	2273 *	–444	.06*	Y	N	.63
Kerala	–12	46	.01	N	A	.07
Maharashtra	3903 *	994	.10	N	I	.61
Madhya Pradesh	4387 *	i19	.09*	N	N	.81
Orissa	3531	–14	.05	Y	N	.51
Punjab and Haryana	1344 *	–355	.10*	Y	N	.93
Rajasthan	5372 *	137	.20	N	N	.59
Tamil Nadu	424	–173	.09*	N	N	.47
Uttar Pradesh	1064*	–268	.05	Y	I	.89
West Bengal	—	—	—	—	—	—

Note : V(1) : Total fertiliser per unit of Net Sown Area
V(2) : Proportion of Canal Irrigation
V(3) : Actual Rainfall

Multicollinearity N No
Y Yes

Auto Correlation N No auto correlation
A Positive significant auto correlation
–A Negative significant auto correlation
I Inconclusive

* Significant at 5 % level.

TABLE 2.6: Multiple Regression Results (OLS)
(Dependent Variable :Foodgrains Volume Index)

Name of the State	Coefficients V(2)	V(4)	V(3)	Multi Collinear -ity	Auto Corre- lation	R^{-2}
Andhra Pradesh	–198	.02*	.02	N	I	.85
Assam	43	.06*	–.002	Y	I	.61
Bihar	–72	.01	.04	N	I	.38
Gujarat	626	.03*	.15*	N	N	.71
Karnataka	–48	.05*	.03*	Y	N	.70
Kerala	15	.03	.01	N	A	.19
Maharashtra	1652	.02*	.07*	N	I	.81
Madhya Pradesh	343	.02*	.07*	Y	I	.78
Orissa	–14	.03*	.05*	Y	N	.66
Punjab and Haryana	–320	.03*	.07	Y	I	.84
Rajasthan	75	.02*	.20*	N	N	.60
Tamil Nadu	–386	.001	.08*	N	N	.35
Uttar Pradesh	–321	.01*	.06*	Y	I	.74
West Bengal	—	—	—	—	—	—

Note : *V*(2) : Proportion of Canal Irrigation
V(3) : Actual Rainfall
V(4) : Proportion of HYV area

Multicollinearity N No
Y Yes

Auto Correlation N No auto correlation
A Positive significant auto correlation
–A Negative significant auto correlation
I Inconclusive

* Significant at 5 % level.

TABLE 2.7: Single Variable Linear Regression Results (OLS)
(Dependent Variable: Agricultural Index)

Name of the States	Coefficients Canal Irrigation	Well Irrigation	Net Irrigation	Forti lizer NPK	HYV area
Andhra Pradesh	2.9* A	3.8*	1.9*	2.2*	.37* A
Assam	–4.6 A	–3.8* A	2.9* A	20.3*	.35*
Bihar	–4.6	–3.7	2.9*	20.3*	.36*
Gujarat	2.6*	.54*	.33*	1.7*	.26* A
Karnataka	2.6*	2.5*	1.4*	1.6* -A	.34*
Kerala	.13 A	.21 A	.15 A	.11 A	.06 A

Contd.)

TABLE 2.7 (*Contd.*)

Name of the States	Coefficients				
	Canal Irrigation	Well Irrigation	Net Irrigation	Ferti-lizer NPK	HYV area
Maharashtra	12.3*	4.6* A	2.7* I	4.1* A	.39* I
Madhya Pradesh	1.1*	.97*	.45*	2.9*	.20*
Orissa	.38	1.2 A	.55*	4.4*	.24*
Punjab and Haryana	2.7* A	1.6* A	1.14* A	2.2* A	.38*
Rajasthan	.94*	.37*	.30*	2.3* I	.15* I
Tamil Nadu	1.3 A	2.8*	1.6* A	2.6*	.02 I
Uttar Pradesh	3.7* I	1.5*	1.1*	3.5* I	.49* A
West Bengal	—	—	—	1.8*	.32*

Note :

Auto Correlation	N	No auto correlation
	A	Positive significant auto correlation
	—A	Negative significant auto correlation
	I	Inconclusive

* Significant at 5 % level.

TABLE 2.8: Single Variable Linear regression Results (OLS) (Dependent Variable : Foodgrain Volume Index)

Name of the States	Coefficients				
	Canal Irrigation	Well Irrigation	Net Irrigation	Ferti-lizer NPK	HYV area
Andhra Pradesh	.17* A	.22*	.11*	.12*	.02* I
Assam	—.65 A	—.54 A	.42* A	.02 A	.05*
Bihar	.09*	.04*	.04*	.13	.01*
Gujarat	.55*	.12*	.07*	.36*	.06*
Karnataka	.34*	.34*	.19*	.21*	.05*
Kerala	.07 A	.11 A	.06 A	.04 A	.04 A
Maharashtra	.83*	.32*	.18*	.24* A	.02* A
Madhya Pradesh	.08*	.08*	.04*	.24* A	.02*
Orissa	.05*	.17*	.07*	.57*	.03*
Punjab and Haryana	.25* I	.15* A	.11* I	.21*	.04* I
Rajasthan	.19*	.07	.06*	.51*	.03*
Tamil Nadu	.09	.09	.09*	.09*	.002
Uttar Pradesh	.07*	.03*	.02*	.07*	.01* I
West Bengal	—	—	—	.09	.02*

Note :

Auto Correlation	N	No auto correlation
	A	Positive significant auto correlation
	—A	Negative significant auto correlation
	I	Inconclusive

* Significant at 5 % level.

TABLE 2.9: Single Variable Autoregressive Linear Regression estimates (Agricultural index as dependent variable)

State / Dependent Variable	Canal Irrigation	Well Irrigation	Net Irrigation	Total Ferti-liser NPK	HYV Area
1. Andhra Pradesh	3.41*	—	—	—	.42*
2. Assam	—.64	3.4	—.58	—	—
3. Bihar	—	—	—	—	—
4. Gujarat	—	—	1.6*	—	.24*
5. Karnataka	—	—	—	1.56*	—
6. Kerala	.81	.11	.62	—.16	—.28
7. Maharashtra	—	2.86*	2.8^{*A}	1.04^{I}	.092
8. Madhya Pradesh	—	—	—	—	—
9. Orissa	—	1.28*	—	—	—
10. Punjab	.016	.22	—.08	.49	—.008
11. Rajasthan	—	—	—	1.80*	.04
12. Tamil Nadu	1.74	—	1.39*	—	.012
13. Uttar Pradesh	3.3*	—	—	3.23*	.48*
14. West Bengal	—	—	—	—	—

* Significant at 5% level
A Positive significant auto correlation
—A Negative significant auto correlation
I Inconclusive

TABLE 2.10: Single Variable Autoregressive Linear Regression estimates (Volume index of foodgrains as dependent variable)

State / Dependent Variable	Canal Irrigation	Well Irrigation	Net Irrigation	Total Fertiliser NPK	HYV Area
1. Andhra Pradesh	.22*	—	—	—	.024*
2. Assam	—.062	.22	—.02	2.73*	—
3. Bihar	—	—	—	—	—
4. Gujarat	—	—	—	—	—
5. Karnataka	—	—	—	—	—
6. Kerala	.013	.033	.01	.12	.002
7. Maharashtra	—	—	—	.10	.002*
8. Madhya Pradesh	—	—	—	.21*	—
9. Orissa	—	—	—	—	—
10. Punjab	—.026	—.058	—.034	—	—.004
11. Rajasthan	—	—	—	—	—
12. Tamil Nadu	—	—	—	—	—
13. Uttar Pradesh	—	—	—	—	08*
14. West Bengal	—	—	—	—	—

*Significant at 5% level
A Positive significant auto correlation
—A Negativesignificant auto correlation
I Inconclusive

Analysis of Inputs

Cross-Section

As we have specified earlier, application of inputs depend upon public investment, private investment, climatic factors and availability of institutional credit, we propose to estimate following linear equations.

(xi) $V7 = f(V8, V9, V5)$
(xii) $V2 = f(V11, V12, V5)$
(xiii) $V2 = f(V11, V7, V5)$
(xiv) $V10 = f(V11, V12, V5)$
(xv) $V10 = f(V11, V7, V5)$

$V2$ = Total fertiliser input (NPK)/Net sown area
$V5$ = Annual rainfall
$V11$ = Crop loan/NSA
$V7$ = Well irrigated area/NSA
$V8$ = Irrigation loan/NSA
$V9$ = Canal Irrigation/NSA
$V10$ = HYV Area/NSA
$V12$ = Total Irrigation/NSA

Annual rainfall has not been any significant determinant for fertiliser input. It is the extent of irrigation either total or well in particular which influences fertiliser use across the region. Well irrigation, however, was not significant in the first phase. It is also found that crop loan and fertiliser use are significantly related across the states in all three phases. Well irrigation, however, has not been explained by irrigation loan. It has rather depended on extent of canal irrigation. Canal irrigation therefore, has either raised the water table through seepage, or demonstrated the effect of irrigation, or induced the cultivator to illegally lift the water from canal through pumpsets. Whatever be it canal irrigation has helped in the growth of well irrigation too. Well irrigation is significantly less in high rainfall areas. Spread of HYV area is again dependent upon extent of irrigation, not on rainfall level. Crop loan has not been important determinant for HYV area.

TABLE 2.11: Multiple Regression Result, Fertiliser Analysis

Independent Variables	Regression Coeffficients					
	Phase 1		Phase 2		Phase 3	
	1	2	1	2	1	2
1. Crop loan/NSA	.029[a]	.028[a]	.017[a]	.019[a]	.010[a]	.012[a]
2. Total Irrigation/NSA	.036[a]	—	.061[a]	—	.147[a]	—
3. Well Irrigation/NSA	—	.050	—	.083[a]	—	.214[a]
4. Rainfall	.1	.03	—.004	.04	—.02	.004
R^{-2}	.79	.63	.89	.77	.84	.71

a = significant at 5% level

TABLE 2.12: Multiple Regression Result, Well Irrigation Analysis

	Regression Coefficients		
	Phase 1	Phase 2	Phase 3
	1	1	1
1. Irrigation loan/NSA	1.581	—.122	—.227
2. Canal Irrigation/NSA	.459[a]	.752[a]	.866[a]
3. Rainfall	—.02[a]	—.03[a]	—.02[a]
R^{-2}	.66	.65	.68

a = significance at 5% level

TABLE 2.13: Multiple Regression Result, HYV Analysis

	Regression Coefficients					
	Phase 1		Phase 2		Phase 3	
	1	2	1	2	1	2
1. Crop loan/NSA	.140	.202	.012	.027	—.010	—.003
2. Total Irrigation NSA	.382	—	.466[a]	—	.635[a]	—
3. Well Irrigation NSA	—	.110	—	.669[a]	—	.923[a]
4. Rainfall	—.3	—.03	—.2	.4	.2	.04
R^{-2}	.20	.04	.76	.50	.80	.55

a = significant at 5% level

Time Series

Application of fertiliser is dependent upon availability of water (either through rain or irrigation), profitability conditions and availability of fund (either from own source or credit). When more land is brought under irrigation, scope of fertiliser (extensive) use increases. Fertiliser use is, however, expected more when extent of well irrigation is more as this assures availability of water. In other words there may be less intensity of fertiliser use when canal irrigation is more dominant. Even otherwise more fertiliser (intensive) may be used for raising the yield. In a particular land how much fertiliser will be used depends upon the input and output prices. Although that is considered an important aspect of fertiliser use that cannot be analysed with present aggregate statistics. Even if the fertiliser application is profitable, the required amount may be used only if enough money is available. This can be possible either from one's own past income and savings or credit available through different institutions. Previous year's agricultural production index has been considered as proxy for agricultural income. This has been adjusted with proper price index to find out the purchasing value of that agricultural production index.

Thus
Index of Agricultural Income(A) = $API \times PIA/PIC$
where API = Agricultural Production Index
PIA = Price Index of Agricultural Commodities
PIC = Price Index of All Commodities
Crop loan (L) is adjusted by price index of chemicals and fertilisers with 1970-71 as base.
Crop loan (L) = crop loan in current price/Price index of chemical and fertiliser

We have estimated two auto regressive linear functions of order one by maximum likelihood estimation procedure.

(xvi) $Ft = f(A_{t-1}, Rt, Wt)$
(xvii) $Ft = f(Ct, Rt, Lt)$

where
Ft = Fertiliser input/Net Sown Area in Year t
A_{t-1} = Index of Agricultural income in year t-1

Rt = Rainfall in year *t*
Wt = Well irrigation/Total irrigation year *t*
Ct = Canal irrigation/Total irrigation in year *t*
Lt = Crop loan in year *t*

Regression co-efficients along with other necessary statistics have been tabulated in Tables 2.14 and 2.15. Using product moment correlation co-efficient (*r*) as the statistic, and .7 (so that r^2 is 0.5) as cut off value (this is definitely neither a scientific nor a satisfactory criterion) multi-collinearity problem seems to be there in case of Maharashtra, Punjab and Haryana, Rajasthan and Uttar Pradesh for first function, and in case of Assam and Punjab and Haryana for second function. Fertiliser use is not very well explained by the variables *A*, *R* and *W* in Assam, Bihar and Kerala. However, when crop loan is used in case of agricultural income, these variables explain the variation well in Kerala. Previous year's income has strong to weak effect on fertiliser use in all states except Andhra Pradesh, Orissa, Rajasthan, Punjab and Tamil Nadu. This, however, does not suggest that income level does not determine the quantum of fertiliser input. It might have been the cumulative effects of past incomes which influenced the fertiliser input in these states. What is, however, clear is that in other states fertiliser input has gone up because of rise in income level. Crop loan, substitute of the present income, however, has been significant (strong to weak) only in Andhra Pradesh, Kerala, Orissa and Punjab. In Tamil Nadu the relationship although weak is found to be negative. Monetary variable, therefore, has been important variable in fertiliser use in all states except Rajasthan and Tamil Nadu. And this monetary variable generally has been the agricultural income, not the credit.

So far as effect of rainfall on fertiliser use is concerned, we get some divergent results from different specification. Combining both the results (Tables 2.14 and 2.15) it seems that only in Andhra Pradesh, Maharashtra and Orissa, fertiliser input has been dependent on rainfall. Andhra Pradesh and Maharashtra among these three states belong to low rainfall region. Considering annual rainfall of 900 mm as a cut off level, other three states which too belong to low rainfall category are Gujarat, Punjab and Rajasthan. As per the first specification, effect of rainfall is significant in Gujarat. According to another specification it is not significant. However, auto correlation seems to exist in that case. Similar situations prevail in Punjab and Rajasthan too. In the light of all these observations, although we cannot firmly put forward, we feel that rainfall influences fertiliser use in low rainfall areas.

TABLE 2.14: Autoregressive Linear Regression Result (Fertiliser Analysis I)

State	Agricultural income	Rainfall	Proportion of well irrigation	$\bar{R}^2$	Auto Correlation	Multi collinearity Product moment correlation coefficient (1,2)	(1,3)	(2,3)
1. Andhra Pradesh	.98E-05	.50[a]E-04	.42[a]	.64	N	.4	.6	.2
2. Assam	.94[b]E-05	.47 E-06	—.34[a]	.32	N	.1	—.03	.4
3. Bihar	.67[c]E-05	.53 E-05	.05	.48	N	.3	.1	.1
4. Gujarat	.346[a]E-04	.201[a]E-04	.06	.73	N	—.01	.6	.4
5. Karnataka	.85[c]E-05	.31E-05	—.83E-02	.80	I[d]	.2	.4	.3
6. Kerala	.77[c]E-04	.33 E-05	.022	.47	N	.4	—.1	.1
7. Maharashtra	.11[a]E-04	.23[a]E-04	—.54[a]	.68	N	.2	.9	.5
8. Madhya Pradesh	.294[c]E-05	.29 E-06	.13[a]	.65	N	—.3	.2	—.04
9. Orissa	.281 E-05	.38[b] E 05	.03[a]	.75	N	—.2	—.04	.1
10. Punjab	.419 E-05	—.19 E-04	.92[a]	.83	I[d]	.3	.9	.4
11. Rajasthan	.412 E-05	.29 E-04	.002	.75	I[d]	.3	.7	.1
12. Tamil Nadu	.357 E -05	.10 E-05	—.06	.82	I[d]	—.4	.5	—.002
13. Uttar Pradesh	.337[a] E-05	.112[b]E-04	.62[a]	.93	I[d]	.03	.7	.03
14. West Bengal	.133[a]E-04	.88[b] E-05	—	.87	I_d	.06	—	—

a = significant at 5%
b = significant at 10%
c = significant at 20%
N = No autocorrelation at 5% level through DW statistic
I = Inconclusive about auto correlation
d = No autocorrelation at 1% level of significance
A = +ve autocorrelation
E = $1\bar{0}$

TABLE 2.15: Autoregressive Linear Regression Result : (Fertiliser Analysis II)

State	Crop Loan	Rainfall	Proportion of canal irrigation	$\bar{R}^{2}$	Auto Correlation	Multi collinearity Product moment correlation coefficient (1,2)	(1,3)	(2,3)
1. Andhra Pradesh	.058[c]	.03[c]E-04	.102	.71	N	.3	.4	—.1
2. Assam	.019	—.12 E-05	—.047	—.11	N	—.6	.8	—.7
3. Bihar	—.008	.79 E-05	.19	.46	N	.01	.4	—.3
4. Gujarat	.012	.8E-05	.03	.67	I	—.4	—.2	—.1
5. Karnataka	.039	.27E-05	.022	.78	N	.5	.4	.1
6. Kerala	.027[a]	—.15 E-05	—.006	.69	N	—.2	.5	—.15
7. Maharashtra	.014	.79[c] E-05	—.17	.65	N	.04	—.06	.3
8. Madhya Pradesh	.010	.98 E-06	.125	.62	N	.2	—.2	.1
9. Orissa	.040[a]	.68[a] E-06	.009[a]	.87	N	—.1	.6	—.2
10. Punjab	.051[b]	.38[c] E-04	—1.35[a]	.86	N	.1	—.8	—.1
11. Rajasthan	.008	.14 E-05	.03[c]	.77	I[d]	.4	—.3	—.04
12. Tamil Nadu	—.017[c]	.19[c]E-06	.96	.85	I[d]	.3	—.1	—.2
13. Uttar Pradesh	.063	.57E-05	—.100	.89	A	—.3	—.2	.1
14. West Bengal	.002	.65 E-05	—	.80	I[d]	.11	—	—

a = significant at 5%
b = significant at 10%
c = significant at 20%
N = No autocorrelation at 5% level through DW statistic
I = Inconclusive about auto correlation
d = No auto correlation at 1% level of significance
A = +ve autocorrelation
E = $\overline{10}$

So far as role of canal irrigation is concerned it has a significant positive effect on fertiliser use only in Orissa. In Punjab and Haryana the co-efficient is negative and significant. However, problem of multi-collinearity seems to exist there. When proporiton of well irrigation increases, it has a significant positive impact on fertiliser in Andhra Pradesh, Madhya Pradesh, Orissa, Punjab and Haryana and Uttar Pradesh. It has a significant negative impact in Assam and Maharashtra. In case of Maharashtra it may be because of problem of multicollinearity. Except for Orissa and Madhya Pradesh other three states Andhra Pradesh, Punjab and Uttar Pradesh belong to group with high to moderate rate of agricultural growth. Other states like Rajasthan, Gujarat, Karnataka and Tamil Nadu belonging to high growth rate zone, however don't show any correlation between well irrigation and fertiliser.

Irrigation

Monetary variable like agricultural income or credit for irrigation, however, determines irrigation level in a different way. It is an investment loan. Well irrigation in a particular year is therefore, dependent on cumulative affect of monetary variables in past years.

Let

Wt= Well irrigated area in year '*t*'
Mt= Monetary variable (income or credit) available in year '*t*' for new irrigation.
B = Input output co-efficient to convert monetary variable to irrigated area

Then

$Wt = Wt_{-1} + BMt$
$Wt_{-1} = Wt_{-2} + BMt_{-1}$
$Wt_{-2} = Wt_{-3} + B\,Mt_{-2}$ and so on

Thus

$$Wt = BMt + BMt_{-1} + BMt_{-2}............ + R \quad (4)$$

Unlike the usual distributed lag model, here, we have constant 'B' co-efficient.

We therefore have two functions

xviii) $W_t = R_1 + \sum_{k=0}^{n} B_1 C_{rt-k}$

xix) $W_t = R_2 + \sum_{r=0}^{n} B_2 A_{t-k}$

where

Crt = Irrigation credit available in year *t*
At = Index of agricultural income in year *t*
n = life of the irrigation implement

TABLE 2.16: Single Variable Autoregressive Linear Regression Results (well irrigation analysis)

	Income			Credit		
	Co-efficient	Auto Correlation	R^{-2}	Co-efficient	Auto Correlation	R^{-2}
	1	2	3	4	5	6
1. Andhra Pradesh	.072[a]	N	.75	.017[c]	N	.75
2. Assam	—.0026	N	.48	—.022[c]	N	.55
3. Bihar	—.021	I[d]	.68	:116[a]	N	.80
4. Gujarat	.32[a]	N	.29	—.086[a]	N	.41
5. Karnataka	.116[a]	N	.52	—.014	N	.04
6. Kerala	—.022	N	.29	.003	N	.28
7. Maharashtra	.044[b]	A	.85	—.023	A	.83
8. Madhya Pradesh	.030	A	90	.006	A	.89
9. Orissa	—.013	N	.34	.087[a]	N	.55
10. Punjab	.116[a]	N	.95	—.002	A	.88
11. Rajasthan	.647[a]	N	.80	.318[a]	N	.92
12. Tamil Nadu	.068[a]	N	.69	—.004	I[d]	.55
13. Uttar Pradesh	.251[a]	N	.89	.034	I[d]	.91
14. West Bengal	—	—	—	—	—	—

a = significant at 5% level
b = significant at 10% level
c = significant at 20% level
N = No auto correlation at 5% level through DW statistic
d = No auto correlation at 1% level of significance
I = Inconclusive about auto correlationA = +ve auto correlation.
A = +vc autocorrelation

Assuming five years of life span, above two functions have been estimated by maximum likelihood method procedure. Agricultural income (as estimated earlier) and irrigation credit (total from all three institutions) have been adjusted to 1970-71 constant price by using price indices of machineries.

What we see from Table 2.16 is that in Andhra Pradesh, Gujarat, Karnataka, Maharashtra, Punjab and Haryana, Rajasthan, Tamil Nadu and Uttar Pradesh well irrigation is explained by agricultural income. All the above states other than Andhra Pradesh and Uttar Pradesh belong to high growth group. On the other hand rural credit explains the variation in well irrigation only in Bihar, Orissa and Rajasthan. Negative effect is found in Gujarat. Whereas crop loan has been instrumental in fertiliser input in four states Andhra Pradesh, Kerala, Orissa and Punjab and Haryana, the irrigation loan has been effective only in three states Bihar, Orissa and Rajasthan.

Summary and Conclusion

What we have thus observed is that whereas technical variables like use of fertiliser, irrigation, HYV seeds etc. or environmental variables like rainfall, soil fertility and economic variable like size of market are responsible for the determination of agricultural performance, the proper environment for the efficient use of such technical inputs depends upon the psychological condition of landlords, public and private investment.

Production relation is definitely an important factor. This, however, need not be understood in the framework of tenancy and semi-feudalism only. Tenancy is not the necessary condition of exploitation and obstacles to agricultural progress. It is the entrepreneurial attitude which determines the agricultural performance. As per this, the entrepreneurial landlords take a long-term view of investment which envisages a larger amount of profit in the long-run, and which calls for unexploitative production relation. Feudal landlords on the otherhand having a myopic view with a rent seeking attitude are interested in the short-run gain, and resort to exploitative measures to derive it. Also, their unentrepreneurial outlook renders the production function inefficient in the sense that inputs do not explain the variations in outputs too well.

What we observe through our empirical exercise is that extent of variation in agricultural output explained by inputs is not of the same order in all states. Whereas more than 75 per cent of variation is explained by inputs in majority of the states, less than 50 per cent of variation is explained in three states viz. Bihar, Kerala and Orissa — all low growth states. Apart from the environmental conditions, psycological conditions of the farmer too are responsible, we feel, for this kind of explanations. Feudalistic farmers tend to be less efficient.

Yield rate has been positively significantly correlated with natural soil fertility. Inspite of use of fertiliser, HYV and irrigation, yield rate has been found to be more where soil fertility is higher. Agricultural production has been found to be significantly dependent on size of market also.

What we have further observed is that irrigation has been major explanatory variable for agricultural growth.[22] While across the region fertilizer growth and agricultural growth have not been commensurate (although coefficients significant in recent phases, R^{-2} is low) fertiliser input has been a significant variable to explain agricultural yield in quite a few states over time.[23] This only suggests that there is unequal pace of fertiliser growth, and because of diminishing returns (may be bacause of intensive use rather than extensive use) agricultural growth does not match with fertiliser growth across the regions.

Although fertiliser use has been found to be correlated with crop loan across the states, rural credit has significant impact on fertiliser use over time in four states only viz. Kerala, Andhra Pradesh, Punjab and Haryana and Orissa. This only suggests that credit has flown more to those areas where fertiliser input was already high, implying in turn whre agricultural income is more. Otherwise fertiliser use is dependent upon well irrigation and agricultural income. In low rainfall areas fertiliser input has been dependent on rainfall. Crop loan has not influenced the spread of HYV area.

So far as well irrigation is concerned, its growth is generally due to the rise in agricultural income. Only in three states viz. Bihar, Orissa and Rajasthan irrigation loans seem to have caused irrigation growth. Among these three Bihar and Orissa are low growth states. Cross sectional analysis suggests that well irrigation has been encouraged by the advent of canal irrigation. Canal irrigation therefore has not only direct but an indirect effect too on agricultural performance over time.

We therefore, finally conclude that agricultural performance has basically been determined by public sector investment specially in canal irrigation, soil fertility, and initial endowment. As a result inputs have been concentrated on high fertility, high income areas. Agricultural credit has not yet played the role it was supposed to in influencing input use, and improving the agricultural performance. Out of the five states viz. Punjab and Haryana, Rajasthan, Andhra Pradesh, Bihar and Orissa where positive impact of credit on either irrigation or fertiliser has been observed, only first three have shown satisfactory performance in agricultural production.

Notes and References

1. Amit Bhaduri (1973).
2. Utsa Patnaik (1986), p. 9
3. Daniel and Alice Thorner (1962), p. 56
4. Kaushik Basu (19840, p. 111
5. Indira Hirway (1986), p. 8
6. Pranab Bardhan (1984), p. 122
7. A. Vaidyanathan (1988), p. 31
8. Mahabub Hossain (1982), p. 154
9. Pranab Bardhan *op. cit.* p. 172
10. Pradhan H. Prasad (1989), p. 84
11. Arun K. Bandyopadhyay (1984) p. 142
12. Pranab Bardhan and Ashok Rudra (1980)
13. Pranab Bardhan *op. cit.* p. 125
14. D.G.M. Newbery (1975)
15. A.K. Ghosh and A. Saith (1976)
16. Pranab Bardhan *op. cit.* p. 125
17. K.N. Raj (1988), p. 267
18. R.P. Kurulkar (1989), p. 42
19. Patil R.K. and Dasgupta R. (1989), p.67
20. Punjab poses a peculiar problem. Ordinary least square estimation method suggest all the co-efficient to be significant but shows positive auto correlation present in each equation. Cochrane-Orcutt method removes auto correlation, but renders estimates insignificant at 5 per cent. Similar results emerge for Maharashtra for fertiliser inputs in case of foodgrain production and HYV seeds in case of agricultural production; for Rajasthan for HYV seeds and for Assam for well irrigation in case of agricultural production.

21. TABLE A: Cross Tabulation of Agricultural and Irrigation growth rates 1966-67 to 1982-83

(Annual Per cent)

Irrigation / Agriculture	High >3	Moderate 2-3	Low < 2
High >4	Gujarat Punjab and Haryana Rajasthan Uttar Pradesh	Maharashtra	
Moderate 2-4	M.P.	Karnataka	Andhra Pradesh Assam
Low <2		Bihar Orissa	Kerala Tamil Nadu

TABLE B: Cross Tabulation of Agricultural and Fertiliser growth rates 1966-67 to 1982-83

(Annual Per cent)

Fertiliser / Agriculture	High >10	Moderate 5-10	Low <5
High <4	Gujarat Punjab & Haryana Rajasthan Uttar Pradesh	Maharashtra	
Moderate 2-4	Andhra Pradesh Karnataka Madhya Pradesh	Assam	
Low <2	Bihar West Bengal	Orissa Tamil Nadu	Kerala

TABLE C: Cross Tabulation of Agricultural and HYV growth rates 1966-67 to 1982-83

(Annual Per cent)

Agriculture / HYV	High>20	Moderate 15 - 20	Low <10
High> 4	Punjab and Haryana Rajasthan	Gujarat Maharashtra	Uttar Pradesh
Moderate 2 - 4	Madhya Pradesh Assam	Andhra Pradesh Karnataka	
Low <2		Bihar Orissa West Bengal	Kerala Tamil Nadu

TABLE D: Cross Tabulation of Foodgrains and Irrigation growth rates 1966-67 to 1982-83

(Annual Per cent)

Food / Irrigation	High > 3	Moderate 2-3	Low <2
High > 4	Gujarat Punjab and Haryana	Maharashtra	
Moderate 2 - 4	Madhya Pradesh Rajasthan Uttar Pradesh	Karnataka	A.P. Assam
Low < 2		Bihar Orissa	Kerala Tamil Nadu

TABLE E: Cross Tabulation of Foodgrain and Fertiliser growth rates 1966-67 to 1982-83

(Annual Per cent)

Food \ Fertiliser	High > 10	Moderate 5 - 10	Low < 5
High >4	Gujarat Punjab and Haryana	Maharashtra	
Moderate 2-4	A.P. Karnataka Madhya Pradesh Rajasthan Uttar Pradesh	Assam	
Low < 2	Bihar West Bengal	Orissa Tamil Nadu	Kerala

TABLE F: Cross Tabulation of Foodgrain and HYV growth rates 1966-67 to 1982-83

(Annual Per cent)

Food \ HYV	High > 20	Moderate 15 - 20	Low < 10
High > 4	Punjab and Haryana	Gujarat Maharashtra	
Moderate 2 - 4	Madhya Pradesh Rajasthan Assam	A.P. Karnataka	Uttar Pradesh
Low < 2		Bihar Orissa West Bengal	Kerala Tamil Nadu

22. The following tables provide R^{-2} of single variable equations estimated with OLS procedure

TABLE G: Dependent Variable (Volume index of foodgrains)

	Canal Irrigation	Well Irrigation	Net Irrigation	Fertiliser	HYV area
Andhra Pradesh	.69	.77	.79	.78	.84
Assam	.14	.002	.35	.16	.66
Bihar	.29	.23	.34	.18	.23
Gujarat	.59	.40	.32	.57	.48
Karnataka	.52	.52	.67	.50	.63
Kerala	.09	.05	.09	.01	.19
Maharashtra	.89	.92	.90	.59	.74
Madhya Pradesh	.48	.47	.46	.55	.52
Orissa	.35	.37	.39	.41	.49
Punjab and Haryana	.56	.84	.82	.87	.83
Rajasthan	.32	.19	.29	.36	.32
Tamil Nadu	.05	.19	.34	.25	.05
Uttar Pradesh	.65	.69	.68	.81	.64
West Bengal	—	—	—	.21	.40

TABLE H : Dependent Variable : (Agricultural Index)

	Canal Irrigation	Well Irrigation	Net Irrigation	Fertiliser	HYV area
Andhra Pradesh	.71	.77	.75	.86	.83
Assam	.16	.002	.37	.60	.69
Bihar	.16	.002	.37	.60	.69
Gujarat	.78	.55	.44	.77	.61
Karnataka	.78	.71	.86	.77	.82
Kerala	.04	.02	.06	.01	.07
Maharashtra	.93	.93	.95	.78	.88
Madhya Pradesh	.40	.39	.38	.44	.45
Orissa	.20	.20	.24	.26	.29
Punjab and Haryana	.58	.84	.83	.86	.86
Rajasthan	.70	.50	.66	.62	.52
Tamil Nadu	.03	.55	.26	.49	.19
Uttar Pradesh	.72	.78	.77	.81	.79
West Bengal	—	—	—	.31	.47

23. See above correlation Tables.

APPENDIX-I

Commercial Bank Rural Credit Advance/outstanding Ratio

	1977	1978	Average
Andhra Pradesh	0.45	0.44	0.445
Assam	0.34	0.31	0.325
Bihar	0.35	0.36	0.355
Gujarat	0.38	0.47	0.425
Karnataka	0.49	0.53	0.510
Kerala	0.83	0.76	0.795
Maharashtra	0.29	0.29	0.290
Madhya Pradesh	0.35	0.33	0.340
Orissa	0.57	0.51	0.540
Punjab	0.49	0.51	0.500
Rajasthan	0.38	0.44	0.410
Tamil Nadu	0.50	0.58	0.540
Uttar Pradesh	0.44	0.45	0.445
West Bengal	0.51	0.41	0.460

Weights of Different Agricultural Items to Compute Agricultural Index

Crop Code	Crop Name	Weight	Crop Code	Crop Name	Weight
01	Rice	33.98	19	Jute	0.81
02	Jowar	4.86	20	Mesta	0.15
03	Bajra	2.83	21	Sugarcane	7.01
04	Maize	3.06	22	Tobacco	1.14
05	Ragi	1.01	23	Potato	1.41
06	Millet	0.81	24	Black Pepper	0.08
07	Wheat	12.16	25	Dry Chillies	1.11
08	Barley	1.34	26	Turmeric	0.31
09	Gram	3.58	27	Arecanut	0.00
10	Tur	1.35	28	Dry ginger	0.07
11	Other Pulses	3.14	29	Banana	1.11
12	Groundnut	4.82	30	Tapioca	0.05
13	Seasum	0.62	31	Cashewnut	0.50
14	Rapeseed & Mustard	1.73	32	Cardamom	0.50
15	Lineseed	0.43	33	Tea	1.85
16	Castorseed	0.11	34	Coffee	0.24
17	Coconut	1.87	35	Rubber	0.19
18	Cotton (lint)	3.01			

APPENDIX-III

Wholesale Price Indices

(1970-71=100)

Year	Chemicals & Products	Machineries	All Commodities	Agricultural Commodities	Food
1966-67	73.8	85.5	80.5	79.6	81.3
1967-68	82.2	90.3	92.7	100.5	102.9
1968-69	87.8	91.1	92.2	98.6	99.3
1969-70	96.0	92.8	94.2	95.1	96.7
1970-71	100.0	100.0	100.0	100.0	100.0
1971-72	101.3	104.6	105.0	99.2	102.1
1972-73	103.9	110.5	113.0	101.3	114.6
1973-74	111.1	118.5	131.6	137.6	135.0
1974-75	156.3	147.4	169.2	164.7	183.6
1975-76	176.8	171.0	175.8	144.4	187.0
1976-77	171.3	170.8	172.4	154.2	150.4
1977-78	172.7	171.1	185.4	182.6	167.2
1978-79	175.6	179.8	185.0	169.0	173.3
1979-80	190.2	207.2	206.5	187.2	179.8
1980-81	232.2	234.1	248.1	210.7	207.3
1981-82	256.9	259.1	278.4	238.4	236.1
1982-83	267.1	275.3	285.3	242.0	242.5
1983-84	278.4	287.4	308.5	268.8	270.4

3

ANALYSIS OF AGRICULTURAL CREDIT

Introduction

As we have observed in the earlier chapter that agricultural credit does not seem to have played it's role, it becomes necessary to have a close look at agricultural credit system. It is more so as rural credit is extremely useful instrument for promoting growth. This promotes agricultural growth directly. This also helps in facilitating economic growth indirectly by (i) increasing purchasing power of poorer people and (ii) moderating the environment of exploitation by strengthening the economic condition of exploited class. These help in increasing the size of the market to sustain a stable growth of both industry and agriculture. This further compels the rent seeker expoliters to invest their money for productive purpose. A well designed agricultural credit policy therefore can take care of both supply and demand side of economic progress.

In this present chapter we intend to analyse the agricultural credit data of (i) scheduled commercial banks including regional rural banks (SCB), (ii) primary agricultural co-operative societies (PACs) and (iii) land development banks (LDB) for last 10-15 years across the states, and other related data to understand the behaviour of agricultural financial institutions, why inspite of substantial growth, agricultural credit has not been able to play it's role.[1]

Data Base

Basic data of agricultural credit has been taken from Report on Currency and Finance (RCF) for commercial banks, and from Statistical Statements Relating to the Co-operative Movement (SSRCM) for primary co-operatives and land development banks. SSRCM provides other data like number of societies, membership etc., also for PACs and LDBs. Other related data like number of offices, deposit level, credit-deposit ratio, etc., for commercial banks have been taken from Banking Statistics : Basic Statistical Returns (BSR) and Statistical Tables relating to Banks in India (STRBI). Land distribution data have been collected from Agricultural Census. Size of cultivators has been taken from Population Census. Yearwise agricultural statistics have been taken from annual reports of Area and Production of Principal Crops, and Statistical Abstracts. Wholesale price indices to convert monetary data at constant prices have been collected from Statistical Abstracts. Fertiliser input data have been taken from Fertiliser Statistics.

There has been a problem of compatibility between credit data of commercial banks on one hand and that of PACs and LDBs on the other hand. Former gives April to March data whereas latter gives July to June data. We have ignored these discrepancies and added yearly disbursement data by each agency to find out total agricultural credit in a year. States have been reorganised as earlier.

Broad Trends

Although there has been a general feeling in the banking circle that rural sector gets more priority so far as the credit deployment is concerned, the data, however, suggests a reverse interpretation. What we find is that credit-deposit ratio for the country as a whole had been 70, 60, and 56 per cent in 1984, 1986 and 1988 respectively, the same ratio for rural and semi-urban areas together being 63, 57 and 52 per cent, respectively in those three years. This only shows that till 1988 rural and semi-urban household finance had been moving out to urban and metropolitan centres. May be the pace has been checked.

Gadgil[2] has already given an indication of growth rate in agricultural lending. Between 1973-74 to 1982-83 although the nominal growth in agricultural lending had been 15 per cent, real growth rate had been 4 per cent only. When, however, we look at growth rates of credit from different institutions, we get a very different picture. What we find from Table 3.1 is that during 1978-87 real growth rate of short-term disbursement (in 1970-71 wholesale price) from commercial banks was 8.2 per cent for all India. This has been positive for all states except for West Bengal where it had been -0.8 per cent. Trend of co-operative credit is however, not as much encouraging.

TABLE 3.1: Real Compound Growth Rate of Disbursement of Rural Credit

(*Percentage per annum*)

Name of the State	Short-Term Credit: Commercial Bank (1978-87)	Short-Term Credit: Primary Agricultural Co-operative Society (1970-85)	Long-Term Credit: Commercial Bank (1978-87)	Long-Term Credit: Land Development Bank (1969-85)
Andhra Pradesh	16.6	7.7	27.6	4.9
Assam	18.1	–8.8	44.8	–33.9
Bihar	7.1	–2.9	16.4	–0.7
Gujarat	10.4	–2.0	26.9	–8.9
Karnataka	11.2	2.34	25.4	0.5
Kerala	14.5	12.6	21.6	13.5
Maharashtra	12.6	0.9	26.6	0.01
Madhya Pradesh	8.8	0.7	18.9	0.78
Orissa	5.4	8.0	20.2	0.16
Punjab	12.5	7.6	17.7	3.8
Rajasthan	11.3	10.7	17.7	5.0
Tamil Nadu	14.3	–0.5	38.2	–8.9
Uttar Pradesh	8.6	3.6	13.9	–0.11
West Bengal	–0.8	7.1	19.1	7.6
All India	8.2	3.9	6.0	2.6

During 1969-70 to 1985-86 real growth rate of co-operative lending for all India had been 3.9 per cent per annum, which is less than half of that of commercial bank short-term advance. Compared to short-term credit, real growth of long-term credit from commercial banks has been 6.0 per cent per annum during 1978-87 whereas similar figure for Land Development Banks (LDBs) had been 2.6 only during 1969-85. It is thus quite clear about the increasing significance of commercial banks lending *vis-a-vis* co-operative lending. However, so far as total agricultural lending is concerned (short and long-term together) co-operative lending is still more. If we break it into short and long-term lending, SCBs have edge over LDBs in long-term lending, and PACS have edge over SCBs in short-term lending (see Table 3.4). What is, however, of concern for co-operative lending for both short and long-term is that, not only these growth rates lag behind those of SCBs, these growth rates are in fact decelerating.

Fitting a regression equation of the type

$$Y_t = B_1 + B_2 t + B_3 t^2$$

where Y_t is advance and t is time, we find from Tables 3.2 and 3.3 that generally B_3 is positive for commercial bank lending and negative for co-operative lending.

Interesting feature about Table 3.4 is that inspite of decelerating trend in co-operative lending, and gaining importance of commercial banks in rural areas, share of credit of SCBs (for both short and long-term) has not shown a secular trend. Rather, it has fluctuated. What is observed is that in bad years, commercial banks' share has significantly reduced. Between 1977-78 to 1984-85, 1979-80 and 1981-82 (see Chapter 1) had been very bad years. On the other hand, other years had been normal to good years. We find a sharp decline in this share in both 1979-80 and 1981-82. This attitude had, however, been across the states. In case of agricultural lending commercial banks therefore seem to be more risk averse compared to co-operative banks.

TABLE 3.2 : Regression Result for Rural Credit at 1970-71 Price as Dependent Variable+

(Commercial Banks)

Name of the State	Dependent Variable = Short-Term Credit Coefficients		Dependent Variable = Long-Term Credit Coefficients	
	t	t^2	t	t^2
Andhra Pradesh	–780.0	152.4**	–1170.7**	147.4*
Assam	7.7	0.4	–216.1*	26.5*
Bihar	–134.8*	14.3*	–1310.7*	149.7*
Gujarat	–210.9**	27.2*	–1181.8	152.0*
Karnataka	–190.3	36.6	–1623.0**	202.5*
Kerala	–474.3	71.1*	–433.0	55.4*
Maharashtra	–159.4	27.8*	–2265.3**	276.6*
Madhya Pradesh	–92.0*	11.2*	–1694.9	202.7*
Orissa	121.5	–9.2	395.7	48.6**
Punjab	–34.9	19.8	–3542.1	433.9**
Rajasthan	–36.2	5.7**	–1527.1**	180.5*
Tamil Nadu	–719.3**	124.7*	–1206.4**	159.9*
Uttar Pradesh	–106.5	16.9	–2613.9**	305.5*
West Bengal	–250.9*	21.9*	–1294.9*	146.8*
All India	–3060.5	521.1*	–20475.8**	2488.4*

+t and t^2 are independent variables

* Significant at 5% level

** Significant at 10% level

TABLE 3.3: Regression Result of Rural Credit Disbursement at (1970-71 Price) as Dependent Variable+

(Co-operative Sector)

Name of the State	Dependent Variable = Short-Term Credit Coefficient	Coefficient	Dependent Variable = Long-Term Credit (1970-85)	(1971-87)
	t	t^2	t	t^2
Andhra Pradesh	98552.5*	–5633.1*	22503.8	–739.4
Assam	–2900.8*	93.6**	813.8	–48.9*
Bihar	9407.7	–899.6*	–6112.5	275.0
Gujarat	69688.8**	–6719.1*	–446222.9*	2173.9*
Karnataka	67953.4*	–4691.9*	–14902.6**	917.9*
Kerala	117477.2**	–5967.4*	4757.1*	67.0
Maharashtra	115721.1	–8519.0	–3926.9	401.9
Madhya Pradesh	29971.2	–2705.6	–217.3	13.4
Orissa	42679.3*	–2222.5**	4221.3**	–235.3
Punjab	171291.4	–9408.3	–20134.9*	1836.1*
Rajasthan	87970.1*	–4684.5*	9617.1*	–412.4*
Tamil Nadu	123883.5*	–8738.5*	–50992.8*	2393.6*
Uttar Pradesh	139511.5*	–8603.8*	2653.4	–232.0
West Bengal	59319.9*	–3396.3*	10775.3*	–575.7*
All India	1130524.0*	–72096.1*	–148264*	14852.4*

+*t* and t^2 are independent variables
* Significant at 5% level
** Significant at 10% level

TABLE 3.4: Distribution of Credit Between Co-operatives and Commercial Banks

(Percentage Share of Commercial Banks)

Name of the State	Short-Term Credit 1977–78	1979–80	1980–81	1981–82	1984–85	Long-Term Credit 1977–78	1979–80	1980–81	1981–82	1984–85
Andhra Pradesh	42.7	34.6	42.1	29.9	57.4	14.6	14.3	22.6	12.9	42.1
Assam	58.8	15.4	22.3	48.3	71.7	57.6	21.3	52.6	—	—
Bihar	23.9	11.2	25.5	9.0	31.8	69.5	41.4	37.9	24.8	52.9
Gujarat	9.5	4.9	15.2	4.0	14.3	70.8	68.7	69.3	50.8	75.3
Karnataka	28.9	16.6	31.9	17.8	32.8	60.6	30.9	55.8	35.6	64.4
Kerala	25.7	9.8	20.4	7.2	16.8	33.4	16.9	44.6	29.9	40.0
Maharashtra	9.4	6.5	4.1	5.9	12.4	41.9	18.6	44.5	24.8	60.7
Madhya Pradesh	6.4	3.8	6.1	3.1	7.6	56.5	49.9	67.9	38.9	64.8
Orissa	18.2	10.1	31.7	7.4	25.9	32.2	15.9	71.9	31.3	65.4
Punjab	8.8	6.3	13.1	5.9	12.5	49.3	41.3	60.6	39.7	55.1
Rajasthan	5.2	2.7	4.4	2.8	7.9	54.3	46.4	60.9	37.4	72.3
Tamil Nadu	27.6	24.5	69.6	31.7	44.5	31.1	57.9	60.1	43.1	67.2
Uttar Pradesh	7.8	4.9	15.7	6.9	14.9	51.5	38.9	63.4	45.2	59.7
West Bengal	22.7	14.7	29.9	15.5	24.1	45.5	36.8	58.0	34.1	60.5
All India	18.5	11.4	18.8	10.6	24.9	45.6	34.4	55.9	34.3	59.0

Credit in Bad Years

This is again supported by the fact that real credit per hectare supplied by commercial banks decline sharply in these bad years. This decline, however, does not seem to be mainly because of one of the expectations as suggested by Binswanger and Khandekar that in a year preceded by a good year, demand of credit declines as result of income effect.[3] This is because same from co-operatives on the other hand does not generally decline so drastically. In fact, at all India level this figure of rural credit per gross cropped areas (Table 3.5) from co-operatives marginally increases in 1979-80 compared to 1977-78. Similar trend has been observed for long-term credit also (Table 3.6).

This decline in the rural credit can not be attributed to lack of demand of credit by cultivators because of bad rain either, as fertiliser use per hectare does not significantly come down in these bad years (Table 3.7). In fact, in 1979-80 which was a very bad year, only 5 states viz., Andhra Pradesh, Assam, Madhya Pradesh, Orissa and Uttar Pradesh experienced reduction in fertilisers input compared to previous year. On the other hand there might have been higher demand of credit for fuels for irrigation purpose.

Coverage of Farmers

Table 3.8 provides per cent of borrowers to total cultivators. Number of cultivators have been obtained from census reports of 1971 and 1981. The compound growth rate computed from these two years has been used to intrapolate the number of cultivators in each year between 1971 and 1981, and extrapolate between 1981 and 1987. What we find is that coverage of cultivators by commercial banks is just about 3 per cent for short-term credit, and around one per cent for long-term credit. Only in 1987, coverage for term lending had increased to about 5 per cent. There seems to be a slow increase in this. Compared to this, coverage by co-operatives is around 20 per cent for short term credit, and this percentage seems to be declining. As a result, total percentage of cultivators borrowing short-term credit has declined in this period. Commercial banks have therefore seem to have only competed with co-operatives by taking their share, not complemented them. So credit widening effect i.e. credit to larger proportion of cultivators across different regions has not materialised by nationalisation of bank, inspite

TABLE 3.5: Short-Term Credit (Rs. 1970-71 Prices) Per Hectare (Gross Cropped) in Different Years

Name of the State	Commercial Bank								Primary Co-operatives							
	1977–78	1979–80	1980–81	1981–82	1982–83	1983–84	1984–85	1985–86	1977–78	1979–80	1980–81	1981–82	1982–83	1983–84	1984–85	1985–86
Andhra Pradesh	29.1	18.9	25.9	15.3	30.9	69.6	66.7	77.5	38.9	35.7	35.7	35.7	—	50.5	49.4	40.0
Assam	1.2	0.4	0.7	0.7	1.9	3.2	2.6	2.7	0.8	2.2	2.3	0.8	—	1.8	1.0	4.4
Bihar	2.9	1.1	3.6	0.9	1.2	1.2	3.7	4.9	9.2	8.9	10.6	10.1	—	8.0	8.0	—
Gujarat	7.6	3.0	7.9	2.7	6.6	9.9	12.0	13.9	72.2	58.2	43.8	64.7	—	72.9	72.0	7.2
Karnataka	15.3	6.9	16.3	8.2	20.4	22.8	23.9	29.3	37.6	34.9	34.8	37.8	—	46.4	49.0	0.5
Kerala	54.7	25.8	66.4	22.7	80.5	90.1	100.3	114.7	158.3	238.3	259.2	293.3	—	453.1	494.8	187.7
Maharashtra	4.5	2.5	5.3	3.3	4.8	7.2	8.3	8.6	43.9	35.1	124.2	52.8	—	53.6	59.1	25.6
Madhya Pradesh	1.1	0.9	1.4	0.7	0.7	1.3	1.7	1.9	16.3	21.8	21.5	23.2	—	22.2	21.0	72.3
Orissa	4.4	3.2	12.8	2.3	8.5	8.9	6.9	4.9	19.9	28.9	27.6	29.3	—	26.2	19.7	—
Punjab	7.0	6.5	13.7	9.9	14.5	16.8	21.0	27.3	73.1	95.8	91.1	158.5	—	188.9	147.5	73.0
Rajasthan	1.0	0.6	0.8	0.8	1.1	1.4	2.1	1.4	18.9	20.3	17.5	25.8	—	21.7	24.0	7.7
Tamil Nadu	39.9	29.7	52.7	37.4	69.9	82.3	109.1	122.5	105.1	91.6	23.1	80.6	—	81.5	135.9	137.6
Uttar Pradesh	2.9	1.9	5.1	2.3	3.0	3.7	5.0	5.5	35.1	35.7	27.4	31.4	—	20.3	28.7	28.2
West Bengal	10.4	3.8	8.5	3.9	5.7	5.7	6.1	10.4	35.4	22.0	19.9	21.5	—	20.6	19.3	0.9
All India	8.9	5.1	10.6	5.4	10.9	15.5	17.3	19.9	39.3	39.9	45.7	45.9	—	51.7	52.2	46.1

TABLE 3.6 : Long-Term Credit (Rs. 1970-71 Price) Per Hectare (Gross Cropped Area)

Name of the State	Commercial Bank								Land Development Bank							
	1977 –78	1979 –80	1980 –81	1981 –82	1982 –83	1983 –84	1984 –85	1985 –86	1977 –78	1979 –80	1980 –81	1981 –82	1982 –83	1983 –84	1984 –85	1985 –86
Andhra Pradesh	3.4	3.4	4.5	2.6	5.9	5.5	14.8	13.4	19.6	20.1	15.5		17.6	18.8	19.4	20.4
Assam	0.8	0.2	0.6	0.4	1.3	1.2	3.4	5.4	0.6	0.8	0.6	—	—	—	—	—
Bihar	5.7	3.7	9.4	3.5	3.7	3.4	10.3	11.3	2.5	5.2	15.4	—	10.6	3.4	5.7	9.1
Gujarat	3.8	4.3	10.4	5.2	9.7	8.9	18.2	16.9	1.6	2.2	4.6	—	4.9	6.7	5.0	5.9
Karnataka	8.8	3.4	4.5	4.9	11.3	10.5	26.3	22.4	5.7	7.5	6.8	—	8.9	11.6	12.5	14.5
Kerala	9.7	6.5	0.6	8.7	17.9	16.5	25.7	23.9	19.3	32.2	24.7	—	20.3	27.1	31.5	38.5
Maharashtra	4.1	2.4	9.4	3.5	5.8	5.4	9.8	13.6	5.7	10.7	7.2	—	10.8	9.4	7.3	6.3
Madhya Pradesh	4.5	3.7	10.4	2.2	4.7	4.4	7.8	7.9	3.4	3.7	3.2	—	3.5	3.9	3.9	4.3
Orissa	2.6	0.9	8.6	2.2	3.8	3.5	4.9	5.5	5.4	4.7	3.9	—	4.7	2.8	2.8	2.6
Punjab	19.5	17.1	19.9	18.4	31.1	28.7	53.9	59.8	20.1	24.3	27.2	—	27.9	29.8	28.3	44.0
Rajasthan	4.7	3.5	5.3	2.6	5.3	4.9	8.1	7.9	3.9	4.0	3.4	—	4.4	3.2	3.2	3.1
Tamil Nadu	4.3	3.2	7.2	6.4	13.8	12.7	42.4	33.0	9.5	2.3	4.7	—	8.5	7.6	17.9	20.7
Uttar Pradesh	8.4	5.5	12.2	6.3	7.4	6.8	12.7	12.6	7.9	8.6	7.0	—	7.7	8.6	8.3	8.6
West Bengal	6.5	1.9	4.4	1.7	2.1	1.9	4.3	17.8	7.7	3.3	3.2	—	3.2	3.3	2.8	2.8
All India	6.3	4.5	10.2	4.6	7.9	7.3	15.1	15.8	7.5	8.5	8.1	—	8.6	8.5	8.9	10.2

TABLE 3.7 : Fertilizer (NPK) Input in Different Years

(*Kg/Hectare*)

Name of the State	1967 –68	1976 –77	1977 –78	1978 –79	1979 –80	1980 –81	1981 –82	1982 –83
Andhra Pradesh	10.4	33.8	41.6	46.0	43.5	46.9	72.3	56.9
Assam	1.8	1.9	3.5	5.6	3.2	3.8	4.7	4.9
Bihar	3.9	13.8	14.9	17.1	17.7	18.4	57.8	33.5
Gujarat	8.7	19.9	28.0	30.9	35.6	33.4	49.3	37.9
Karnataka	5.5	20.9	24.4	32.4	32.9	32.2	43.8	36.0
Kerala	18.9	23.6	26.8	34.6	37.0	34.1	36.4	38.4
Maharashtra	9.1	14.7	18.2	19.2	21.0	20.8	32.9	25.2
Madhya Pradesh	1.1	3.5	7.4	8.9	7.7	9.2	14.9	8.7
Orissa	2.8	8.6	8.1	8.7	8.3	8.7	12.8	14.9
Punjab	1.4	33.4	54.3	65.7	78.9	80.9	124.5	127.7
Rajasthan	1.5	5.8	6.7	7.6	8.9	7.8	9.6	9.0
Tamil Nadu	21.1	39.4	55.6	62.1	70.5	77.2	102.9	76.6
Uttar Pradesh	8.6	31.5	36.8	43.5	42.7	46.8	76.9	65.9
West Bengal	6.7	19.9	21.8	30.9	32.4	37.1	46.3	37.4
All India	7.2	20.4	25.5	29.2	31.0	31.8	48.4	37.3

of high growth of commercial banks' rural credit. Nationalisation of bank was, although expected to help progress in the direction of socio-economic democracy.[4]

Statewise experience is, however, different although in almost all the states this coverage is quite low. States like Kerala, Punjab and Tamil Nadu experienced a very high coverage. This coverage has, however, declined in 1986-87. Interesting point to note about Kerala is that this coverage is more than 100 per cent in both 1980-81 and 1986-87 implying that a cultivator has taken loan twice in a year whereas at all India level hardly 25 per cent of cultivators seemed to have availed of formal credit. This total coverage in case of short-term credit has increased only in four States Andhra Pradesh, Gujarat, Maharashtra and Madhya Pradesh in 1986-87 over 1980-81.

We have got correlation coefficient between coverage by SCBs and PACs for short-term, and between that by SCBs and LDBs for long-term credit (Table 3.9). Except in 1980-81, for short-term credit, all others show positive correlation indicating that commercial banks have gone for lending in those areas only where co-operatives have already gone belying the hope that SCBs would complement co-operatives, not compete with them.

TABLE 3.8 : Percentage of Borrower to Total Cultivators in Different Years

Name of the State	(1980-81) Short-Term SCB	PACS	Total	Long-Term SCB	LDB	Total	(1986-87) Short-Term SCB	PACS	Total	Long-Term SCB	LDB	Total
Andhra Pradesh	5.14	21.0	26.14	0.49	1.55	2.04	9.6	20.4	30.0	1.6	—	—
Assam	—	—	—	—	—	—	—	—	—	—	—	—
Bihar	0.52	23.8	24.32	0.37	0.44	0.81	0.27	11.5	11.77	0.44	—	—
Gujarat	1.20	15.10	17.30	0.59	0.15	0.74	1.7	25.1	26.8	0.93	—	—
Karnataka	3.8	13.1	16.9	0.65	0.98+	1.63	4.8	0.8*	5.6	2.0	—	—
Kerala	43.3	167.9	221.2	4.1	5.1	9.2	50.6	81.5**	132.1	7.4	—	—
Maharashtra	1.0	15.6	16.6	0.5	0.99	1.49	1.5	15.7	17.2	4.4	—	—
Madhya Pradesh	0.29	11.90	12.99	0.21	0.25	0.46	0.42	21.60	22.02	0.67	—	—
Orissa	4.9	21.4	26.3	1.3	0.67	1.97	1.7	13.5	15.2	0.86	—	—
Punjab	4.2	55.5	59.7	1.9	2.2	4.1	3.4	25.6***	29.0	9.3****	—	—
Rajasthan	0.30	19.9	20.2	0.61	0.41	1.02	0.35	6.5++	6.85	0.52	—	—
Tamil Nadu	11.9	43.1	60.0	0.44	0.06	0.50	14.4	17.1	31.5	7.0++++	—	—
Uttar Pradesh	0.98	16.5	17.5	0.42	0.28	0.70	0.93	5.5+++	6.43	0.53	—	—
West Bengal	1.8	14.7	16.5	0.37	0.32	0.69	2.1	3.9	6.0	0.71	—	—
All India	2.7	21.2	23.9	0.61	0.54	1.15	3.4	14.9	18.3	1.0	—	—

* This was extraordinary low in 1985-86. In 1984-85 it was 11.5.

** Till 1985-86 there was monotonic increase, and it was 314.4 in 1985-86

*** 49.1 in 1985-86

**** 2.4 in 1985-86

\+ 1979-80

++ 20.5 in 85-86

+++ 14.4 in 1985-86

++++ 2.1 in 1985-86

TABLE 3.9: Correlation Coefficient Between Coverage By Commercial Banks, and PACS and LDBs

	1980-81		1985-86	
	PAC	LDB	PAC	LDB
1. Commercial Bank (Short-Term)	–0.17	–	0.60	–
2. Commercial Bank (Long-Term)	–	0.10	–	0.70

Distribution of Credit

One important aspect of rural credit is its distribution among different size classes. Regional Rural Banks had been specifically established for catering weaker sections. It is otherwise generally believed that in any market economy dominant section will be able to mop up a larger share of institutional credit offered at a concessional rate. We have the share of credit by small and marginal farmers (upto 2 hectares) from different credit institution (Table 3.10). In 1969-70, when commercial bank was not much present in the scene of agricultural credit, small and marginal farmers had a share of 26.9 per cent of short-term and 24.8 per cent of long-term credit which had been provided by PACs and LDBs respectively (Table 3.11). Within the country, different states however had different experiences. It (short-term) was as low as 15.7 per cent in Madhya Pradesh and as high as 62.3 per cent in Kerala. This variation may be because of local socio-political condition or land distribution pattern. According to 1980-81 census small and marginal farmers of Kerala have access to 63.6 per cent of land whereas those in Madhya Pradesh have access to 12.4 per cent of land only. What is however of importance is that with the emergence of commercial banks as important lenders in rural areas, and that too with a bias in favour of small and marginal farmers, these farmers have been able to evince a larger share of credit from both PACs and LDBs too. As a result share of total short-term (SCB and PACs) or long-term (SCB and LDB) credit availed by small and marginal farmers seems to have increased.

TABLE 3.10: Share of Short-Term Credit by Small and Marginal Farmers

Name of the State	1969-70			1979-80			1982-83			1985-86		
	SCB	PACS	Total	SCB	PACS	Total	SCB	PACS	Total	SCB	PACS	Total
Andhra Pradesh	—	37.1	37.1	53.6	41.0	50.1	59.4	35.6	47.8	67.4	48.9	60.4
Assam	—	40.2	40.2	88.0	70.9	79.3	66.1	75.5	68.6	74.4	90.3	78.6
Bihar	—	48.6	48.6	64.5	22.5	27.2	77.0	39.1	43.5	75.0	37.9	49.0
Gujarat	—	16.4	16.4	22.1	17.7	17.9	20.1	18.1	18.8	23.4	19.2	20.2
Karnataka	—	14.8	14.8	42.2	32.5	35.7	51.1	41.8	45.4	57.4	30.8	41.1
Kerala	—	62.3	62.3	81.3	59.6	66.1	90.9	65.0	71.3	91.7	60.6	66.7
Maharashtra	—	20.2	20.2	31.0	18.2	19.3	36.2	25.5	26.5	36.2	28.7	29.7
Madhya Pradesh	—	15.7*	15.7*	32.1	20.4	20.9	38.9	24.7	24.7	41.8	30.2	31.6
Orissa	—	44.5*	44.5*	86.0	47.9	51.7	43.0	62.1	57.3	74.6	39.0	48.8
Punjab	—	25.2	25.2	32.0	25.6	26.8	38.1	26.2	27.4	38.0	24.9	27.0
Rajasthan	—	21.3	21.3	21.0	20.5	20.9	19.5	28.8	28.4	31.8	40.9	40.4
Tamil Nadu	—	25.4	25.4	76.7	18.4	33.3	80.0	46.3	63.1	78.6	41.7	59.7
Uttar Pradesh	—	33.1	33.1	45.2	51.2	51.0	52.6	48.4	48.8	58.6	52.8	53.9
West Bengal	—	25.1**	25.1**	72.5	63.3	65.0	75.0	80.2	79.0	70.0	67.1	73.7
All India	—	26.9	26.9	56.4	32.6	36.4	61.7	37.0	42.3	64.8	38.9	46.4

* for 1970-71
** for 1973-74

TABLE 3.11: Share of Long-Term Credit by Small and Marginal Farmers

Name of the State	1969-70			1979-80			1980-81		
	SCB	LDB	Total	SCB	LDB	Total	SCB	LDB	Total
Andhra Pradesh	—	29.2	29.2	34.4	—	—	59.9	—	—
Assam	—	86.6	86.6	64.0	89.7	82.0	38.2	95.2	53.7
Bihar	—	46.5	46.5	52.1	39.2	44.5	55.6	87.2	76.6
Gujarat	—	9.1*	9.1*	18.4	—	—	12.0	5.4	10.5
Karnataka	—	21.9	21.9	23.6	—	—	21.6	—	—
Kerala	—	98.7*	98.7*	62.7	87.7	83.5	48.0	—	—
Maharashtra	—	20.8+	20.8+	26.0	31.1	30.1	24.9	34.1	29.5
Madhya Pradesh	—	7.1	7.1	30.5	—	—	28.1	—	—
Orissa	—	—	—	72.9	—	—	60.5	—	—
Punjab	—	12.3	12.3	8.7	19.9	12.8	8.9	—	—
Rajasthan	—	8.9	8.9	21.1	32.2	27.0	26.4	25.5	26.1
Tamil Nadu	—	53.1	53.1	33.7	49.1	33.9	42.7	54.2	47.3
Uttar Pradesh	—	38.8	38.8	29.1	50.9	42.5	20.9	41.2	35.8
West Bengal	—	—	—	47.8	80.7	68.4	51.6	86.6	66.1
All India	—	24.8	24.8	25.1	43.4	34.5	25.4	48.8	31.1

* 1968-69
+ 1971-72

Desai[5] and Gadgil,[6] specially latter are perhaps little concerned about this increase as Gadgil writes "the small farms have received more than their *fair* share" (*emphasis added*). A further inquiry, however, is required to analyse this distribution. If we look at the share of net irrigated area by small and marginal farmers (Table 3.12) share of credit availed by smaller farms does not seem to be that much fair. If we look at the share of wholly irrigated area, and if we assume that institutional credit generally has flown to these areas only, smaller farmers, we think, have yet some scope to demand for a little more fairer share of credit. Although small and marginal farmers have 44 per cent of wholly irrigated area (1980-81 census), they enjoy 42 per cent of short-run credit (1982-83) only.

The problem, however, we think is much more than this. If we look at the distribution of operational holdings, the share of small farms in institutional credit is not only not fair but very much unfair. Since institutional credit covers up to 15 per cent of cost only,[7] farmers specially smaller ones require much more fund than what they get. Whereas large farmers because of their larger income have surplus fund with them to invest in agriculture, smaller cultivators have in fact negative surplus fund with them to invest in agriculture. A micro-level study at Jalgaon district[8] (Table 3.13) shows that small farmers because of low level of income have negative surplus with them (consumption cannot be pushed below a certain floor level) for agricultural operation. Although the finding is from a small region, and that too in a drought year, this indicates a general pattern. If we look at the NCAER[9] study of 1968-69, 1969-70, and 1970-71, similar pattern is observed for all India also (Table 3.14). Surplus is negative for smaller farmers. But considering the fact that repayment of debt for household expenditure (mainly from non-institutional sources) has not been explicitly included in the consumption menu (it might have been taken as miscellaneous item), the consumption of the smaller farmers seems to have been underestimated, and hence the surplus overestimated.

TABLE 3.12 : Share of Different Resources by Small and Marginal Farmers (Per cent)

Name of the States	Short-Term Credit		Long-Term Credit	Holding of Land	Culti-vated Area	Net Irri-gated Area	Wholly Irrigated Area
	1979-80	1982-83	1979-80	1980-81	1980-81	1980-81	1980-81
Andhra Pradesh	50.1	47.8	—	78.0	30.0	—	—
Assam	79.3	68.6	82.0	82.1	36.8	—	—
Bihar	27.2	43.5	44.5	86.5	41.6	—	—
Gujarat	17.9	18.8	—	45.9	12.9	—	—
Karnataka	35.7	45.4	—	59.1	19.4	—	—
Kerala	66.1	71.3	83.5	96.1	63.6	—	—
Maharashtra	19.3	26.5	30.1	51.9	12.4	—	—
Madhya Pradesh	20.9	24.7	—	50.5	15.5	—	—
Orissa	51.7	57.3	—	73.6	37.7	—	—
Punjab	26.8	27.1	12.8	15.1	11.7		
Rajasthan	20.9	28.4	27.0	48.9	9.5	—	—
Tamil Nadu	33.3	63.1	33.9	86.5	46.9	—	—
Uttar Pradesh	51.0	48.8	42.5	86.8	48.3	—	—
West Bengal	65.0	79.0	68.4	89.2	60.4	—	—
All India	36.4	42.3	34.5	74.4	26.2	36.7	44.1

TABLE 3.13: Household Surplus (Income-Consumption) of Rural Households in Girna Command Area at Jalgaon District, 1987

(Rupees per capita annual)

Water Regime	Occupational Group					
	Landless	Artisans	Small Farmers	Middle Farmers	Big Farmers	Large Farmers
Head	–316.3	–298.3	–397.8	93.8	976.4	1880.1
Middle	–575.7	–719.8	–810.9	410.1	873.6	1245.1
Tail	–321.3	87.5	–432.7	–165.0	556.0	1910.2
Non Command	–387.7	–607.1	–662.5	–164.1	2388.7	1426.1

Source : Patil and Dasgupta, 1989.

TABLE 3.14 : Household Surplus (Income-Consumption)

(Annual Per Household in Rupees)

Land Size in Hectares	1968-69	1969-70	1970-71
Nil	353	191	276
Below 1	–58	–26	–72
1.1 – 2.5	–60	60	22
2.6 – 4.5	426	627	734
4.6 – 6.5	1212	1182	1078
6.6 – 8.5	245	836	1763
8.6 – 10.5	1355	2022	553
10.6 – 14.5	2460	5091	3877
14.5 and above	2040	6506	7807
All Classes	294	368	437

Note: Computed from Tables 3.4 and 3.24 of Changes in Rural Income in India, NCAER, 1975.

This is again indirectly corraborated from all India debt and investment survey data of 1981-82.[10] Total borrowing from different sources, borrowings at differet rate of interest, and borrowings for different purposes have been given according to different asset size group. If distribution of cultivators according to asset group, and distribution of households according to land size, are compared, a correlation between asset size and land size is obtained, so that one can be taken as a proxy for other (Table 3.15). Asset group up to Rs. 50,000 can be considered to constitute marginal and small farmers with land up to 2 hectares; Asset group of Rs 50,000 – 1,00,000 constitutes semi-medium farmers with land size 2 to 4 hectares, and asset group Rs 1,00,00 and above may be considered to constitute medium and large farmers with land size more than 4 hectares. What is now evident from Table 3.16 is that 34 per cent of the borrowing by small farmers is for household consumption. Since this is hardly available through institutional sources, they depend heavily from non-institutional sources at a higher rate of interest (above 15 per cent).

TABLE 3.15 : Distribution of Cultivators, and Agricultural Holdings, All India

Asset Size* Group (Rupees)	No. of* Cultivators (00)	Distribution* of Land (Percentage)	Size Group** (Hectare)	No. of Holding** (000)	Distribution of holding** (Per cent)
Upto Rs 1,000	11,276	1.6			
			upto 1 (Marginal)	50,122	56.4
1,000 – 5,000	80,684	11.3			
			1 - 2 (Small)	16,072	18.1
5,000 – 10,000	95,989	13.4			
10,000 – 20,000	1,48,223	20.7			
20,000 – 50,000	2,06,590	28.9			
50,000 – 1,00,000	99,963	14.0	2 - 4 (Semi-medium)	12,455	14.0
1,00,000 – 5,00,000	69,604	9.7	4 - 10 (medium)	8,068	9.1
5,00,000 and above	3,204	0.4	Above 10 (Large)	2,166	2.4
Total	7,15,534	100	Total	88,883	100

* All India Debt and Investment Survey, 1981-82

** Agricultural Census, 1980-81.

Large farmers on the other hand can meet their demand of credit mostly from institutional sources at much concessional rates. Although large farmers also borrow for household expenditure (11 per cent of total borrowing), this does not seem to be because of negative savings. It is primarily to avail concessional credit which may be used directly for lending at a very high rate. On the contrary small farmers are in genuine need of much larger credit at concessional rate. If we have a mechanistic view as share of credit to be proportional to operated or even irrigated area, we are perhaps going to confound the inbuilt inequality because of land distibution. Large farmers have already larger share of land because of historical reasons, and hence larger concentration of wealth. Now if they are given concessional credit according to holding size, credit institutions, we think, are increasing the inequality against the planning goals. On the other hand credit must be need-based, and this need should include minimum household nutritional and other requirements too. Otherwise the generated income out of credit (it is assumed that credit will not be misused as it is given in kinds, not in cash) will be diverted to repay the debt from non-institutional source provided at a very high rate of interest, and institutions will face recovery problem. This is however not a new concept as more than two decades back Gadgil[11] had proposed full credit covering consumption and production needs. If there is a fund constraint, the total credit may be diverted to small farmers only. Large farmers may be asked to finance the whole business out of their own generated income. This will, we think, achieve a higher agricultural growth.

TABLE 3.16: Pattern of Borrowing

	Percentage of* Borrowing for Household Expenditure Borrowing	Percentageof** Institutional Borrowing Interest	Percentage of*** Borrowing Below 15% of Rate of to Total
Marginal and Small Farmer (upto 2 hectare)	34	40	46
Medium Farmer (2-4 hectare)	21	53	52
Large Farmer (4 hectares and above)	11	72	63

Computed from All India Debt and Investment Survey 1981-82

* Rest of the borrowing for non-household expenditure.

** Rest of the borrowing from non-institutional sources.

*** Rest of the borrowing at interest rate higher than 15 per cent.

If cropping intensity be any criterion to evaluate the performance, bigger farmers do not seem to have done better (See Table 3.17). If large farmers have become more efficient compared to small farmers in post green revolution period, it is mainly because of their comparatively greater access to concessional institutional finance. Micro-level study at Jalgaon,[12] however, suggests that although return on expenditure is not higher for small farmers (Table 3.18), return on land is definitely higher for small farmers, atleast on irrigated holdings (Table 3.19). This implies that small farmers are efficient so far as yield is concerned. They however, achieve this at higher cost, may be because of higher cost of their fund. If it is assumed that credit is available mainly for irrigated lands, and the fact that return on land is higher for small farmers in irrigated land, and reverse is the case for unirrigated land (Table 3.19) where credit is not generally available, it can be inferred that credit is more efficiently used by small farmers, and it may be expected to be true even when credit is available for unirrigated land. It is therefore, important to revise the agricultural lending policy for both growth and distibution.

TABLE 3.17: Cropping Intensity for Cultivators of Different Landholding Size, 1980-81

Name of the State	Marginal	Small	Semi-Medium	Medium	Large	All
Andhra Pradesh	1.22	1.17	1.19	1.13	0.94	1.14
Assam	1.28	1.19	1.26	1.31	1.09	1.24
Bihar	1.42	1.33	1.45	1.62	1.33	1.45
Gujarat	1.06	1.07	1.06	1.06	1.04	1.06
Haryana	1.56	1.53	1.55	1.54	1.54	1.54
Himachal Pradesh	1.78	1.71	1.63	1.54	1.42	1.64
Jammu and Kashmir	1.32	1.38	1.46	1.47	1.40	1.39
Karnataka	1.35	1.10	1.08	1.05	1.04	1.07
Madhya Pradesh	1.24	1.15	1.16	1.14	1.12	1.14
Maharashtra	1.14	1.11	1.10	1.09	1.07	1.09
Punjab	1.40	1.59	1.71	1.76	1.68	1.71
Rajasthan	1.39	1.26	1.23	1.17	1.06	1.14
Tamil Nadu	1.24	1.22	1.31	1.28	1.17	1.25
Uttar Pradesh	1.46	1.39	1.45	1.41	1.32	1.42
Remaining States and UTs	1.34	1.28	1.24	1.18	1.09	1.25
All India	1.34	1.25	1.25	1.21	1.12	1.22

Source: Agricultural Census 1980-81

TABLE 3.18: Return on Expenditure for Different Size Groups of Farmers at Jalgaon District, 1987

	Small Farmer	Middle Farmer	Big Farmer	Large Farmer
Head				
Irrigated	2.88	2.62	1.70	2.99
Unirrigated	1.62	2.05	2.01	2.41
Middle				
Irrigated	1.36	1.00	2.77	1.84
Unirrigated	1.22	4.27	2.66	2.71
Tail				
Irrigated	2.30	–0.11	1.52	2.04
Unirrigated	2.26	2.63	2.57	2.45
Non-Command				
Irrigated	4.05	1.26	2.20	1.42
Unirrigated	1.27	1.19	1.65	2.62

Source: Patil and Dasgupta (1989).

TABLE 3.19: Return from Land for Different Size Group of Farmers at Jalgaon District, 1987

(*Rupees per Hectare*)

	Small Farmer	Middle Farmer	Big Farmer	Large Farmer
Head				
Irrigated	11257	6284	3529	4479
Unirrigated	2169	2289	2278	2405
Middle				
Irrigated	2566	795	1149	1236
Unirrigated	1467	3771	2481	2361
Tail				
Irrigated	4135	–30	1812	2636
Unirrigated	1741	2246	1923	2101
Non-Command				
Irrigated	12797	2016	3661	2485
Unirrigated	1435	1397	1695	1831

Source: Patil and Dasgupta (1989).

Another way of understanding the distribution of credit is to find out what percentage of cultivators in a particular size group has been covered by either commercial banks or PACs. At all India level, it seems that percentage of cultivators (holdings as proxy) of small landholdings are less covered by CBs and PACs. Considering the fact that percentage of farmers who have outstanding debt with commercial banks is also quite small, commercial banks do not seem to have penetrated much into the interior of rural economy. For PACs, coverage of farmers, in fact, seem to be declining. Since co-operative banking sector is the major instrument for disbursing short-term production credit in rural areas, and as that sector seems to be shrinking (decelerating growth), total percentage coverage has declined. This is applicable to both small and non-small farmers. Another point to be noted here is that except in Maharashtra and Gujarat, proportion of non-small farmers covered by institutional finance is more than that for small farmers in all other states. Also considering the well established fact of "leakages", and coverage of very small segment of small and marginal farmers, it seems to be unfair to assert that the resource is being tilted in favour of smaller farmers. On the contrary, it may be possible that many of these small farmers are representatives of bigger ones, and hence quite influential to cover a larger share. It goes with the middle class ideology of urban bankers of maintaining a *status quo* and siding with dominant sections of rural areas. Also banks seem to have been satisfied with the target of "forty per cent" of lending to priority sector in general and agricultural sector in particular; not by penetrating into the needy sector. Not that there has not been any growth in the number of borrowers, it has not been matched with the growth of cultivators.

So far as the bias towards non-small farmers by co-operative sectors is concerned, it can be either at the very first stage that is at the entry point, or at the actual disbursement level. Membership itself may be difficult to obtain, or even after getting membership, concessional credit may be cornered away by non small farmers. What is evident (Table 3.23) that there is not much difference in percentage of members availing credit across the size class. But when we look at the membership position (Table 3.24), we find that a very small percentage of small farmers are members to these societies, whereas a much larger segment of non-small farmers are members of the society. It therefore seems that membership drive is a necessary condition for poorer farmers to get co-operative credit. This however, may not be a suffi-

TABLE 3.20: Percentage of Borrowers to Total Holdings for short-term credit (Commercial Banks)

Name of the State	Small Farmers (<2 hectares)						Non-Small Farmers (>2 hectares)					
	1980–81	1981–82	1982–83	1983–84	1984–85	1985–86	1980–81	1981–82	1982–83	1983–84	1984–85	1985–86
Andhra Pradesh	4.9	2.7	4.6	12.7	11.6	8.7	5.0	2.6	4.5	8.8	8.7	8.0
Assam	0.5	0.2	0.3	0.5	0.6	0.5	0.2	0.1	0.5	0.4	0.6	0.4
Bihar	0.4	0.19	0.17	0.28	0.17	0.18	0.62	0.29	0.20	0.32	0.27	0.33
Gujarat	1.56	0.78	1.26	1.60	1.89	2.05	1.84	0.06	1.41	1.91	2.30	2.61
Karnataka	5.79	2.09	5.22	5.31	5.82	6.59	3.06	1.34	3.02	3.22	4.13	4.93
Kerala	9.23	2.16	7.46	7.51	6.47	6.93	8.54	1.82	10.66	10.99	12.04	11.18
Maharashtra	1.22	0.58	0.93	1.54	1.49	1.86	1.37	0.72	1.18	1.52	1.79	1.65
Madhya Pradesh	0.43	0.08	0.33	0.64	0.81	0.79	0.85	0.58	0.34	0.60	0.66	0.62
Orissa	5.23	1.49	1.52	1.78	3.09	2.65	8.13	0.51	2.20	2.16	1.92	1.12
Punjab	6.82	4.22	4.61	5.14	5.47	8.36	7.08	2.63	5.20	5.89	5.99	6.77
Rajasthan	0.32	0.20	0.21	0.33	0.50	0.42	0.54	0.28	0.47	0.73	0.64	0.60
Tamil Nadu	9.50	3.71	7.20	8.34	8.82	9.25	6.34	2.95	8.12	9.47	10.76	12.19
Uttar Pradesh	0.86	0.43	0.45	0.59	0.82	0.89	2.23	1.22	1.19	1.42	1.64	1.83
West Bengal	1.49	0.68	1.48	1.13	1.08	1.59	0.96	0.31	0.40	0.55	0.48	1.10
All India	2.96	1.18	2.30	3.30	3.32	3.26	2.52	1.06	1.94	2.62	2.83	2.92

TABLE 3.21: Percentage of Borrowers to Total Holdings for short-term credit (PACS)

Name of the State	Small Farmers (<2 hectares)						Non-Small Farmers (>2 hectares)					
	1980 –81	1981 –82	1982 –83	1983 –84	1984 –85	1985 –86	1980 –81	1981 –82	1982 –83	1983 –84	1984 –85	1985 –86
Andhra Pradesh	11.42	13.03	10.64	10.12	13.76	13.04	33.52	27.9	28.6	37.4	28.8	26.6
Assam	1.33	1.51	1.04	1.41	7.96	7.82	0.81	1.78	1.45	1.45	—	—
Bihar	7.66	10.13	12.11	11.59	3.55	2.99	62.44	77.24	98.00	98.92	48.10	42.19
Gujarat	23.88	21.39	20.36	22.25	29.82	31.03	24.84	23.04	24.59	22.58	30.76	25.26
Karnataka	12.14	13.89	13.47	10.57	10.98	9.07	16.73	14.12	12.72	15.32	16.39	17.17
Kerala	23.14	21.59	22.01	23.20	31.61	34.69	228.22	226.51	244.21	272.56	146.76	170.35
Maharashtra	12.81	15.44	19.32	20.19	21.28	21.36	22.97	24.25	21.96	19.93	18.99	19.89
Madhya Pradesh	12.98	14.49	16.49	17.40	16.20	15.78	20.51	24.02	22.18	21.07	23.06	22.74
Orissa	20.44	25.70	14.54	26.43	21.34	11.55	36.94	25.34	29.57	24.45	23.88	33.43
Punjab	77.43	81.60	80.17	66.35	100.94	65.95	91.66	86.69	63.91	82.37	93.01	87.65
Rajasthan	21.73	25.87	31.33	29.49	28.70	29.69	32.42	29.48	34.51	31.38	29.36	27.17
Tamil Nadu	26.78	23.91	8.70	0.04	6.62	7.75	87.07	76.01	48.31	0.21	42.67	46.73
Uttar Pradesh	13.67	12.19	10.47	11.55	11.28	11.09	54.19	45.64	56.12	44.83	44.30	44.55
West Bengal	12.49	10.77	10.76	5.96	6.49	7.25	30.17	10.26	13.02	19.97	20.25	13.06
All India	15.38	15.64	13.89	13.07	13.98	13.47	37.96	35.78	36.18	34.17	31.95	31.17

TABLE 3.22 : Percentage of Total Borrowers of Short-Term Credit to Total Holdings

Name of the State	1980 –81	1981 –82	1982 –83	1983 –84	1984 –85	1985 –86	1986 –87
Andhra Pradesh	23.03	20.46	20.83	30.39	32.96	29.78	29.85
Assam	1.64	1.71	1.46	1.86	7.04	6.86	—
Bihar	15.50	19.06	23.04	22.42	8.87	8.62	—
Gujarat	26.11	22.99	24.03	24.30	32.94	30.94	36.68
Karnataka	18.70	15.77	17.53	16.97	20.24	20.02	6.12
Kerala	40.34	31.36	37.49	39.32	47.98	52.64	18.10
Maharashtra	19.08	20.47	21.80	21.86	22.54	22.96	20.24
Madhya Pradesh	17.22	19.52	19.55	19.80	20.41	20.01	36.50
Orissa	30.78	26.89	20.40	27.98	25.08	20.22	—
Punjab	92.30	87.91	76.24	81.52	125.47	98.66	55.10
Rajasthan	27.62	27.96	33.32	36.05	30.20	29.65	10.03
Tamil Nadu	43.96	34.33	21.02	8.61	21.08	23.20	23.41
Uttar Pradesh	20.04	17.06	16.82	16.41	16.37	16.26	6.81
West Bengal	15.82	11.39	12.43	8.42	9.35	9.77	4.63
All India	26.86	23.07	23.93	24.57	26.14	25.52	21.08

TABLE 3.23: Percentage of Borrowing Members to Total Members in Co-operative Sectors 1981-82

Name of the State	Small Farmers		Non-Small Farmers 2 Hectares or More
	Up to 1 Hectare	1-2 Hectare	
Andhra Pradesh	23.9	27.1	25.1
Assam	2.0	1.6	6.7
Bihar	20.6	48.9	48.2
Gujarat	48.4	46.6	40.9
Karnataka	27.9	23.9	17.3
Kerala	36.2	35.5	81.9
Maharashtra	28.2	29.3	30.3
Madhya Pradesh	33.4	32.6	34.4
Orissa	40.8	39.0	33.6
Punjab	73.9	73.6	80.3
Rajasthan	48.1	43.8	39.1
Tamil Nadu	44.2	72.6	56.6
Uttar Pradesh	39.2	40.1	38.6
West Bengal	28.6	46.7	14.2
All India	33.3	38.6	38.9

cient condition as non-small farmers may even then tilt the resource in their favour by means of their bargaining power. Small farmers need to be organised to get their share.

TABLE 3.24: Percentage of Members of Co-operative Sectors to Total Landholdings 1981-82

Name of the State	Small Farmers		Non-Small Farmers
	Up to 1 Hectare	1-2 Hectare	2 Hectares or More
Andhra Pradesh	38.4	75.2	111.2
Assam	62.0	128.8	26.7
Bihar	17.6	115.5	160.2
Gujarat	44.7	45.2	56.3
Karnataka	45.3	65.2	81.4
Kerala	42.1	303.1	276.6
Maharashtra	45.1	64.1	79.9
Madhya Pradesh	28.8	69.8	69.8
Orissa	54.6	81.5	75.5
Punjab	91.0	136.2	107.9
Rajasthan	43.4	76.4	75.4
Tamil Nadu	36.9	77.9	134.3
Uttar Pradesh	21.6	71.1	118.1
West Bengal	23.2	55.5	72.5
All India	31.6	81.4	91.9

Recovery Rate

The problem of recovery has always been there, and it creates a recycling problem. This becomes more severe for co-operative sectors as they depend upon local resource whereas commercial banks may have little more flexibility because of their national character. It is easy to calculate recovery rate for co-operative sector as they provide data on "recovery" and "overdue" in a particular year. This implies —

$$\text{demand} = \text{recovery} + \text{overdue}$$

and recovery rate can be given by

$$\text{recovery rate} = \text{demand/recovery} \times 100$$

This is, however, not so easy for commercial banking sector as we do not have above type of data. We have rather only "disbursement" and "outstanding" data.

Let O_t = outstanding at the end of the year *t*

L_t = Loan issued during the year *t*

D_t = Demand raised during the year *t*

R_t = Amount recovered during the year *t*

r = Average rate of interest on the outstanding amount (including fresh loan)

RR_t = Recovery Rate

Assuming that short-term credit should have been repayed within a year (after the harvesting), we expect following relationship to hold approximately.

$$D_t = O_{t-1} * (1 + r) + L_t/2 * (1 + r)$$

$$R_t = O_{t-1} * (1 + r) + L_t/2 * (1 + r)$$

$$+ L_t/2 * (1 + r/2) - O_t$$

$$= D_t + L_t/2 * (1 + r/2) - O_t$$

Hence,

$$RR_t = (R_t/D_t) * 100$$

What has been assumed here is that demand for repayment is raised against whole of previous year's outstanding and only half of the total advance of current year along with interest rate. Demand for repayment of another half of the loan which might have been disbursed in the later half of the year would be raised in the following year. Average rate of interest has been considered as 11 per cent.[13]

What we observe (see Tables 3.25, 3.26, 3.28 and 3.29) is that recovery rate for small farmers is not necessarily poorer than that of non-small farmers as expected because of higher cost of fund for that group. At all India level, recovery rate for smaller farmers for cooperative sector had in fact been at least as good as that of non-small farmers till 1982-83. Only in 1983-84, recovery rate for non-small farmers had been little better. For commercial banks, although recovery rate for large farmers was much better in earlier years, there is not much difference in the recovery rates of small farmers and non-small farmers in recent years. The point to be noted, however, is that recovery rate of commercial banks in recent years is much lower at about 40 per cent compared to 60 per cent of co-operatives. And recovery rate of

TABLE 3.25 : Recovery Rate (PACS)

Name of the State	1977-78			1980-81			1982-83			1983-84		
	SF	NSF	Total	SF	NSF	Total	SF	NSF	Total	SF	NSF	Total
Andhra Pradesh	67.1	65.3	67.0	56.4	55.8	56.1	61.4	65.2	63.7	53.3	59.9	57.5
Assam	20.9	99.6	99.6	15.9	26.9	16.9	11.0	9.3	11.3	11.7	9.8	11.7
Bihar	—	—	—	45.9	31.8	36.8	35.5	31.8	33.0	27.8	23.9	25.2
Gujarat	55.6	56.2	56.1	58.2	52.9	54.2	60.2	63.2	62.5	53.9	64.7	62.7
Karnataka	57.6	57.7	58.4	49.5	51.1	51.5	51.1	53.5	53.2	61.2	66.2	64.9
Kerala	71.4	68.4	70.2	80.9	76.1	79.6	82.2	74.4	79.9	82.3	73.9	79.8
Maharashtra	54.5	54.5	54.6	48.0	47.9	48.0	58.0	58.5	58.0	59.0	59.4	59.2
Madhya Pradesh	43.6	45.7	45.4	47.7	50.9	50.3	52.1	55.3	54.5	52.6	55.7	54.9
Orissa	42.1	48.8	45.9	66.9	53.8	60.3	60.2	57.8	59.2	54.5	56.3	55.2
Punjab	63.9	66.2	65.4	72.9	73.9	73.4	69.1	78.4	75.8	75.5	80.3	78.5
Rajasthan	—	100.0	100.0	52.5	57.3	56.2	57.2	63.2	63.2	58.3	63.2	61.7
Tamil Nadu	68.1	64.4	67.1	22.6	29.1	26.6	41.7	41.9	41.9	49.0	47.3	48.1
Uttar Pradesh	55.5	46.0	51.6	62.2	58.4	60.2	54.7	48.2	51.5	49.8	50.6	50.2
West Bengal	—	—	—	36.9	43.2	38.7	49.0	23.7	41.1	30.8	36.5	32.4
All India	60.6	58.2	59.3	55.2	54.1	54.6	59.4	59.9	60.1	58.0	61.4	60.7

SF : Small Farmers
NSF : Non-small Farmers

TABLE 3.26 : Recovery Rate (Commercial Bank, Short-Term Credit)

Name of the State	1980-81				1982-83				1986-87			
	Up to 1	1-2	Above 2	Total	Up to 1	1-2	Above 2	Total	Up to 1	1-2	Above 2	Total
Andhra Pradesh	37.2	35.4	71.5	50.9	44.9	41.6	52.1	46.5	44.1	48.1	44.9	45.9
Assam	12.7	24.7	85.2	39.2	37.9	–1.9	38.9	28.6	55.9	24.7	26.2	40.9
Bihar	3.7	11.1	54.2	22.5	1.2	12.7	3.7	6.3	19.1	48.5	28.5	32.6
Gujarat	34.7	42.9	95.8	83.6	31.1	26.4	55.5	49.0	52.9	64.5	58.8	59.3
Karnataka	52.0	41.4	93.9	72.5	51.2	33.9	51.7	46.7	38.9	34.9	41.5	38.9
Kerala	54.2	48.4	97.4	59.5	45.9	43.6	37.1	44.5	55.6	59.0	13.2	51.4
Maharashtra	56.7	27.9	86.1	72.4	41.7	15.9	22.9	24.6	40.2	29.5	40.3	38.0
Madhya Pradesh	18.0	16.4	82.6	61.1	38.2	12.8	28.6	26.9	46.2	39.9	45.8	44.3
Orissa	75.4	87.3	113.7	89.8	22.3	41.5	37.9	35.5	39.6	45.9	29.1	39.7
Punjab	90.4	70.0	104.5	97.4	43.1	55.2	51.1	51.1	20.5	33.9	52.7	45.2
Rajasthan	55.7	50.9	99.9	90.4	29.2	9.1	42.9	36.6	19.8	56.1	62.5	55.9
Tamil Nadu	62.3	53.2	81.9	65.1	38.8	41.9	40.1	40.0	53.8	47.5	50.4	50.9
Uttar Pradesh	32.8	36.1	101.9	71.9	35.7	31.5	47.8	39.8	43.8	51.2	54.6	50.5
West Bengal	45.2	34.6	55.2	43.9	16.5	7.7	54.9	25.6	30.2	26.5	72.0	43.7
All India	51.8	42.3	86.6	64.2	39.9	36.4	44.8	40.9	46.8	45.6	47.3	46.7

co-operatives is quite stable compared to that of commercial bank. Rath's[14] recovery rate with original data (supplied by NABARD), however, shows a better recovery rate for commercial banks (Table 3.27). Reserve Bank of India[15] too finds a higher recovery rate for commercial banks. Neither the basis of calculation nor the nature of data, however, has been given in either of these studies. Our estimates change if we vary rate of interest and distribution of credit between different seasons,[16] but conclusion regarding repayment behaviour remains same.

Credit and Area Covered

Purpose of short-term loan has been to facilitate the use of fertiliser and other inputs for agricultural production. In the post green revolution period agricultural activity has become cost intensive, and that is why small farmers have gradually fallen into disadvantageous situation.

TABLE 3.27: Overdues as Per cent of Annual Demand for Repayment of Short Term Loans by Commercial Banks, 1979-85

Name of the State	1979-80	1980-81	1981-82	1982-83	1983-84	1984-85
Haryana	26.3	33.3	35.9	33.7	35.8	36.1
Himachal Pradesh	55.3	56.6	62.3	66.2	54.9	56.4
Jammu and Kashmir	63.6	60.0	57.4	58.0	65.8	40.4
Punjab	17.6	21.0	22.8	22.3	36.5	23.0
Rajasthan	37.1	94.0	40.4	38.4	41.2	43.0
Assam	86.4	75.7	69.7	62.1	55.8	80.5
Bihar	70.2	69.2	62.7	66.9	23.6	67.6
Orissa	58.6	66.8	60.3	61.7	59.6	51.5
West Bengal	64.8	64.4	65.6	67.2	58.6	63.6
Madhya Pradesh	56.4	54.8	50.3	52.3	53.6	50.8
Uttar Pradesh	45.1	41.4	46.2	41.8	41.9	43.9
Gujarat	48.5	46.2	44.7	47.0	46.2	46.0
Maharashtra	44.1	38.3	42.2	42.2	49.0	46.5
Andhra Pradesh	43.6	42.8	41.5	40.1	41.1	39.4
Karnataka	41.0	40.7	40.5	38.6	44.3	45.1
Kerala	42.7	26.9	30.4	26.6	25.2	22.6
Tamil Nadu	41.9	36.2	37.7	36.5	38.1	32.9
Others	—	—	45.1	52.5	41.7	41.6
India	42.8	47.0	41.5	41.2	52.5	39.8

Source: Rath, N. (1987)

TABLE 3.28: Recovery Rate (Co-operative Sector) All India Short-Term Credit

Year	Up to 1 hectare	1-2 hectares	2-4 hectares	4-8 hectares	Above 8 hectares	Tenant Cultivation	Agricultural Labourer	Others	Total
1977-78	61.8	59.7	58.5	58.4	57.7	66.6	74.9	66.8	59.3
1978-79	60.5	58.9	55.9	56.7	57.5	70.9	72.1	67.5	58.1
1979-80	65.5	58.5	59.5	53.4	58.2	62.9	72.7	61.5	58.8
1980-81	56.0	54.6	57.1	49.3	55.7	55.1	62.9	58.6	54.6
1981-82	61.1	60.9	60.7	57.5	59.7	60.4	65.0	72.6	60.2
1982-83	63.1	56.2	57.9	57.9	66.2	63.3	63.9	76.8	60.1
1983-84	58.9	57.4	59.9	59.3	66.7	67.2	73.3	68.6	60.7

TABLE 3.29 Recovery Rate, (Short-term), All India Commercial Banks

Year	Up to 1 hectare	1-2 hectares	Above 2 hectares	Total
1980-81	51.82	42.34	56.59	64.25
1981-82	16.33	10.41	−13.78	4.84
1982-83	34.18	25.86	36.15	32.99
1983-84	39.94	36.44	44.77	40.87
1984-85	43.20	41.72	38.53	41.06
1985-86	44.24	40.62	38.31	40.97
1986-87	46.84	45.58	47.35	46.65

We have tried to estimate the extent of land brought under fertilisers use through financial credit. It is assumed that short-term credit taken by the cultivators has indeed been used for fertiliser application.

We formulate the following relation—

$$Aijt = Pijt * Lijt * Ui$$
$$= (Bijt * Lijt * Ui)/LHijt$$

where for *i*th class of farmer in *j*th state in the year *t*

Aijt = Area under credit
Pijt = Proportion of borrowers to total number of landholdings
Bijt = Number of borrowers
Lijt = Area of total holdings
LHijt = Number of landholdings
Ui = Proportion of total landholding area used for cultivation by *i*th class of farmers at all India in 1980-81.

Time series data of borrowers (*Bijt*) is available in Report of Currency and Finance. Data on number of holdings and area of holdings for different class is available for only two census years 1970-71 and 1980-81. For each class and each state growth rate has been computed between these two years, and that has been used to intrapolate data between 1970-71 and 1980-81, and also to extrapolate beyond 1980-81 up to 1986-87. For intrapolation, 1970-71 data have been taken as base, and for extrapolation 1980-81 data have been taken as base.

TABLE 3.30 : Area Under Short-Term Credit Commercial Bank

('000 ha)

Name of the State	1979-80			1982-83			1986-87		
	SF	NSF	Total	SF	NSF	Total	SF	NSF	Total
Andhra Pradesh	118.9	376.9	495.9	211.7	408.4	620.1	529.1	930.6	1459.0
Assam	4.2	1.6	–5.8	5.2	14.5	19.7	8.1	13.0	21.1
Bihar	12.5	27.1	39.6	7.9	11.5	19.4	10.3	16.6	26.9
Gujarat	7.4	52.9	60.3	16.7	114.2	130.9	30.5	228.1	258.6
Karnataka	41.5	86.2	127.7	120.0	263.2	383.2	151.8	340.3	492.1
Kerala	43.4	32.2	75.6	86.5	63.9	150.4	89.6	62.6	152.2
Maharashtra	17.5	94.7	112.2	33.1	196.6	229.7	82.5	272.9	355.4
Madhya Pradesh	5.5	80.4	85.9	9.3	59.7	69.0	26.7	147.9	174.6
Orissa	18.8	16.9	35.7	27.8	64.9	92.7	49.5	39.3	88.8
Punjab	25.5	172.3	197.9	37.6	316.3	353.9	53.6	373.7	427.3
Rajasthan	3.2	47.3	50.5	4.1	76.8	80.9	12.5	153.7	166.2
Tamil Nadu	161.5	155.5	317.0	261.6	295.8	557.4	376.6	505.5	882.2
Uttar Pradesh	39.5	116.5	156.0	38.5	99.3	137.8	76.2	155.8	231.9
West Bengal	30.7	16.9	47.6	51.0	7.8	58.8	69.9	16.5	86.4
All India	557.6	1367.9	1925.6	974.0	2116.5	3090.6	1608.3	3444.7	5053.0

SF = Small farmers with operational holding less than 2 hectares.
NSF = Non-small farmers with operational holding more than 2 hectares.

TABLE 3.31 : Area Under Short-Term Credit (Co-operative Societies)

('000 ha)

Name of the State	1979-80			1982-83			1986-87		
	SF	NSF	Total	SF	NSF	Total	SF	NSF	Total
Andhra Pradesh	441.9	3679.3	4121.2	485.2	2608.3	3093.3	756.4	1931.9	2688.3
Assam	9.1	27.0	36.1	18.1	40.5	58.6	141.8*	—	141.8*
Bihar	343.8	4511.3	4855.1	570.8	5612.9	6183.7	156.1	2290.2	2446.2
Gujarat	208.3	2445.9	2654.2	270.4	1990.8	2261.2	346.0	2873.2	3219.2
Karnataka	265.1	1919.6	2184.7	309.6	1107.1	1416.7	224.7*	1495.2*	1719.9
Kerala	240.7	1646.2	1886.9	255.2	1464.9	1720.1	433.8*	1015.1*	1448.9
Maharashtra	517.4	3768.0	4285.4	684.9	3642.8	4327.7	697.5	3328.5	4026.0
Madhya Pradesh	256.6	3569.2	3825.8	456.7	3929.6	4386.3	477.1*	4049.8*	4526.9
Orissa	322.1	960.0	1282.1	266.6	870.8	1137.4	200.1	950.9	1151.0
Punjab	489.9	5169.2	5659.1	654.7	3887.2	4541.9	521.9*	5362.7*	5884.6*
Rajasthan	354.1	5302.1	5656.2	603.4	5681.2	6284.6	619.2*	4426.9*	5046.0
Tamil Nadu	901.2	3477.8	4379.0	316.3	1760.8	2077.1	204.6	1782.9	1987.6
Uttar Pradesh	1117.2	3562.5	4679.7	896.1	4684.5	5580.7	988.9*	3590.7*	4579.7
West Bengal	488.2	1032.8	1521.0	370.8	254.6	625.4	276.8*	243.3*	520.1
All India	5881.5	43184.6	49066.1	5683.5	37049.1	42732.6	6189.2*	33614.6*	39803.8

* 1985-86

TABLE 3.32 : Irrigated Area, All India

('000 hectares)

Year	Net Irrigated Area	Gross Irrigated Area
1950-51	20,853	22,563
1955-56	22,758	25,642
1960-61	24,661	27,980
1965-66	26,344	30,901
1970-71	31,103	38,195
1975-76	34,491	43,189
1980-81	38,805	49,585
1982-83	40,721	52,121
1983-84	41,955	53,937
1984-85	41,779	54,067

TABLE 3.33: Short Term Credit (1970-71 Price) Per Hectare (area under credit) 1985-86

(Rupees/hectare)

Name of the State	Commercial Banks		Co-operative Sector	
	Below 2 hectares	Above 2 hectares	Below 2 hectares	Above 2 hectares
Andhra Pradesh	1534.5	469.3	447.4	114.6
Assam	1030.1	325.4	31.5	—
Bihar	3982.1	703.1	158.5	19.9
Gujarat	1175.7	540.3	313.0	280.9
Karnataka	1208.7	341.9	749.3	238.7
Kerala	3659.0	431.6	1978.2	449.8
Maharashtra	1054.1	422.2	380.4	254.4
Madhya Pradesh	800.9	244.3	295.8	79.1
Orissa	708.7	348.4	343.2	100.2
Punjab	1590.8	414.5	675.7	158.3
Rajasthan	981.0	190.3	291.8	55.5
Tamil Nadu	1710.3	395.0	1140.4	275.9
Uttar Pradesh	1064.4	403.5	377.9	92.4
West Bengal	881.9	1117.3	370.9	121.4
All India	1525.9	394.1	550.8	145.0

Commercial banks have increased their operation from 1.9 million hectares to just 5 million hectares. Co-operatives have on the other hand reduced their operation from 49 million to 40 million hectares. As a result total area under institutional credit has declined from 50 to 45 million hectares (Table 3.30 and 3.31). Ironically, these figures are very close to irrigation figure (Table 3.32) which tends to confirm the general feeling that institutional credit is confined to irrigated areas only.

Another interesting point to be noted is that area covered under credit by both commercial banks and co-operative societies are more for non-small farmers (>2 hectares). However, unit credit (credit per hectare) available to these non-small farmers is much less. As we have seen earlier that these non-small farmers generally have internal resource to finance the agricultural activity, additional resources if diverted to small farmers, will either reduce their financial dependence on non-institutional sources, and hence improve the recovery of institutional finance, or can bring additional area under credit.

$$\text{Additional area} = \frac{\text{(Credit disbursed to non-small farmers)}}{\text{(Unit credit for small farmers)}}$$

If that kind of strategy be adopted, additional 11 million hectares of land can be brought under credit network (Table 3.34), of course under the assumption that non-small farmers will be able to carry out their agricultural activity without institutional support.

TABLE 3.34: Additional Area which can be Brought Under Institutional Finance without Additional Resource 1985-86

('000 ha)

Name of the State	Short-Term Commercial Bank	Short-Term Co-operatives	Long-Term Commercial Bank
Andhra Pradesh	221.4	612.3	64.7
Assam	3.4	15.3	8.6
Bihar	3.2	292.6	11.0
Gujarat	97.3	1841.8	61.9
Karnataka	121.5	476.3	75.1
Kerala	7.8	230.8	5.6
Maharashtra	110.4	2220.6	97.7
Madhya Pradesh	33.6	1083.1	60.6
Orissa	15.6	277.7	10.8
Punjab	107.8	1256.5	71.9
Rajasthan	19.0	841.2	75.9
Tamil Nadu	97.9	393.2	35.6
Uttar Pradesh	55.9	878.2	71.6
West Bengal	26.1	79.7	64.1
All India	920.9	10484.0	715.1

What we have thus so far observed is that commercial banks have not been able to penetrate into agricultural economy significantly. There has however, been significant increase in rural and semi-urban bank branches. We have an idea about the increase in rural and semi-branches from Table 3.35.

TABLE 3.35 : Increase in Rural and Semi-urban Branches of Commercial Banks

Name of the State	Number		Density (Number/hectare)	
	1979-80	1986-87	1978-79	1986-87
Andhra Pradesh	1392	2893	0.0607	0.1262
Assam	537	1299	0.0307	0.0842
Bihar	1134	3640	0.0776	0.2475
Gujarat	1016	1732	0.0666	0.1136
Karnataka	1296	2460	0.0757	0.1439
Kerala	1322	1860	0.3730	0.5387
Maharashtra	1317	2662	0.0464	0.0938
Madhya Pradesh	1025	3003	0.0258	0.0758
Orissa	552	1492	0.0377	0.1019
Punjab	1423	2117	0.1683	0.2495
Rajasthan	788	1996	0.0264	0.0669
Tamil Nadu	1163	2125	0.1085	0.1992
Uttar Pradesh	1996	5532	0.0758	0.2104
West Bengal	860	2098	0.1158	0.2823
All India	15821	34909	0.0617	0.1373

Expansion, Recovery and Credit

What is generally expected is that with branches moving into interiors of rural society, the bank personnel will have better understanding of the area and more utilisation of the credit. So with the expansion of bank branches and hence increase in the density of rural and semi-urban offices, credit deployment per hectare, recovery rate and savings mobilisation are expected to increase.

We have computed correlation between recovery rate and both rural offices per hectare and rural and semi-urban offices together per hectare. Former variable gives a better correlation.

What we find from Table 3.36 is that there is very weak correlation between branch penetration and recovery rate. The states with higher density of rural branches do not necessarily have a better recovery rate. What is, however, interesting to find is that these correlations are higher (although marginally) for cultivators with less than one hectare

TABLE 3.36: Correlation Co-efficient between Density of Credit Outlet and Recovery Rate for Different Classes of Farmers

Year	Commercial Bank (Rural Offices/hectare)			Co operatives (Societies/hectare)	
	Less than 1 hectare	1-2 hactares	More than 2 hectares	Less than 2 hectares	More than 2 hectares
1977-78	–	–	–	–0.03	–0.48
1978-89	0.25	0.28	0.19	–0.11	–0.22
1979-80	0.32	0.22	0.19	0.18	–0.31
1980-81	0.47	0.41	0.32	0.10	0.11
1981-82	0.30	0.14	–0.01	0.34	0.02
1982-83	–0.11	0.48	0.01	0.28	0.02
1983-84	0.26	0.31	0.23	–	–
1984-85	–	–	–	–	–
1985-86	0.34	–0.05	0.11	–	–

TABLE 3.37: Correlation Co-efficient Between Density of Rural Offices and Recovery Rate Over Time (1979-87)

Name of the State	Commercial Bank				Co-operative Sectors		
	Below 1 hectare	1-2 hectares	Above 2 hectares	Total	Below 1 hectare	Above 2 hecrares	Total
Andhra Pradesh	0.20	0.02	0.00	0.04	0.47	0.30	0.46
Assam	0.28	–0.29	0.02	0.25	0.72	–0.71	–0.62
Bihar	0.32	0.04	–0.17	–0.03	–0.60	–0.59	–0.60
Gujarat	0.19	0.74	0.19	0.30	0.12	–0.21	–0.18
Karnataka	0.16	–0.35	–0.01	–0.04	0.81	0.68	0.82
Kerala	–0.16	–0.18	–0.12	–0.27	–0.66	–0.89	–0.73
Maharashtra	0.25	–0.34	0.01	0.00	0.94	0.84	0.88
Madhya Pradesh	0.35	–0.09	–0.05	0.01	–0.52	–0.57	–0.57
Orissa	0.36	0.30	–0.10	0.27	–0.51	–0.94	–0.81
Punjab	–0.03	–0.04	0.02	–0.06	–0.63	–0.83	–0.95
Rajasthan	0.14	–0.26	0.19	0.06	0.82	–0.80	–0.79
Tamil Nadu	0.18	–0.17	0.04	0.08	–0.44	–0.45	–0.43
Uttar Pradesh	0.30	0.45	0.21	0.36	0.19	0.59	0.44
West Bengal	0.22	–0.16	0.50	0.45	0.19	0.13	0.86

of land, or with 1 to 2 hectares of land compared to non-small farmers. Branch density therefore has impact on recovery, if any, for small and marginal farmers only. The above pattern is again confirmed if we look at time series data of each state, and we compute correlation co-efficient between density of offices and recovery rate. Table 3.37 suggests that most of the correlation co-efficients are quite low rejecting the hypothesis of an association between branch expansion and recovery rate. Whatever value of coefficients are there in Tables 3.36 and 3.37, they do suggest that branch expansion if at all help in recovery, has helped in the recovery from small and marginal farmers only. A moderate to good correlation is observed for density of co-operative society in Andhra Pradesh, Karnataka, Maharashtra, Uttar Pradesh and West Bengal.

If the branch expansion of commerical banks had taken place with the objective of providing credit it is expected that states with higher branch density will be providing a higher level of credit per hectare (total gross cropped area) since their area of operation will increase. Also, within a state credit per hectare over time is supposed to increase with branch expansion.

TABLE 3.38: Correlation Co-efficient between Density of Rural Branches (Rural branch/hectare) and Credit Per Hectare (1970-71 Price) for Commercial Banks

Year	Short-term Credit	Long-Term Credit
1979-80	0.61	0.53
1980-81	0.74	0.62
1981-82	0.48	0.61
1982-83	0.62	0.58
1983-84	0.42	0.58
1984-85	0.40	0.57
1985-86	—	—
1986-87	0.21	0.53

Although unit credit seems to be more in the states where rural branch density is more (Table 3.38), degree of this positive correlation has been declining overtime, and in recent years this correlation has been quite weak. This correlation is, however, little higher for long-term unit credit, although this too is declining. This can be further understood if we concentrate on each state and look at time series data of branch density and unit credit in rural areas. We find a very weak relationship for short-term credit (Table 3.39). It is in fact negative in some states. Correlation between density and unit long-term credit is, however, generally quite higher. Considering the fact that the long-term credit is mainly enjoyed by non-small farmers, it appears that branch expansion has tried to serve non-small farmers of local areas. On the other hand, relationship between branch expansion and deposit mobilisation is almost perfect (Table 3.40). Even the cross sectional relationship is quite close (Table 3.41).

TABLE 3.39: Correlation Co-efficient between Branch Density and Credit Per Hectare (1970-71 Price), Over Time (1979-87) for Commercial Banks

Name of the State	Short-Term Credit	Long-Term Credit
Andhra Pradesh	0.17	0.51
Assam	0.27	0.66
Bihar	–0.05	0.58
Gujarat	0.03	0.48
Karnataka	0.03	0.47
Kerala	–0.47	–0.15
Maharashtra	0.30	0.47
Madhya Pradesh	0.31	0.55
Orissa	0.31	0.50
Punjab	–0.19	0.30
Rajasthan	0.62	0.63
Tamil Nadu	0.19	0.46
Uttar Pradesh	0.23	0.60
West Bengal	–0.04	0.54

TABLE 3.40: Correlation Co-efficient between Density of Rural Branches and Deposit Per Hectare (1970-71 Prices)

Name of the State	Current	Savings	Term	Total
Andhra Pradesh	0.99	0.98	0.98	0.99
Assam	0.90	0.99	0.99	0.99
Bihar	0.95	0.94	0.95	0.95
Gujarat	0.99	0.98	0.99	0.99
Karnataka	0.93	0.98	1.00	1.00
Kerala	0.94	0.98	0.95	0.97
Maharashtra	0.97	0.99	0.99	1.00
Madhya Pradesh	0.91	0.99	1.00	1.00
Orissa	0.88	0.99	0.98	0.99
Punjab	0.93	0.97	0.91	0.94
Rajasthan	0.87	0.97	0.98	0.98
Tamil Nadu	0.99	0.99	0.97	0.98
Uttar Pradesh	0.96	0.98	1.00	0.99
West Bengal	0.98	1.00	1.00	1.00

It therefore appears that commercial banks have penetrated into the rural areas not with the motive of supplying credit for agriculture but for deposit mobilisation. Former aspect seems to be the byproduct of later aspect, and credit has flown to irrigated areas with low risk. Binswanger and Khandekar[17] with district level data too observed that banks and credit institutions have tended to locate and grow more rapidly in districts with high agroclimatic potential, and in the regions where road densities and regulated market network have been increasing. They further maintained that these institutions avoided areas with high risks from either flood, a short rainy season, low moisture capacity or lack of irrigation. As a result, the regional inequality in terms of credit deployment in rural areas which was supposed to have been reduced because of entry of commercial banks in agricultural sector has not taken place. Whereas inequality in branch distribution has decreased (Table 3.42) i.e. more branches might have been opened in non-banked areas, credit inequality for both long-term and short-term has increased over time. Inequality in deposit mobilisation however, seems to have been marginally reduced or at best has remained constant, if we look at share of deposit by different states for different years (Table 3.44). It therefore, shows that branch expansion indeed has been utilised mainly for deposit mobilisation.

TABLE 3.41: Correlation Co-efficient between Density of Rural Branches and Deposit Per Hectare (1970-71 Prices)

Year	Current	Savings	Term	Total
1979-80	0.35	0.96	0.92	0.95
1980-81	0.27	0.96	0.90	0.95
1981-82	0.46	0.96	0.92	0.96
1982-83	0.56	0.91	0.80	0.88
1983-84	0.48	0.90	0.80	0.86
1984-85	0.48	0.88	0.73	0.83
1985-86	—	—	—	—
1986-87	0.50	0.83	0.68	0.78

TABLE 3.42: Regional Inequality of Rural Credit (Gini Co-efficients)

Year	Short-Term Credit			Long-Term Credit			
	Commer-cial Bank	Co-opera-tive Society	Total	Commer-cial Bank	Land Develop-ment Bank	Total	Commer-cial Bank Rural Branch
	(1)	(2)	(3)	(4)	(5)	(6)	(7)
1977-78	0.49	0.22	0.26	0.29	0.34	0.22	—
1978-79	0.49	0.24	0.28	0.31	0.10	0.31	—
1979-80	0.55	0.26	0.27	0.34	0.33	0.32	0.075
1980-81	0.44	0.22	0.26	0.32	0.27	0.24	0.065
1981-82	0.54	0.21	0.23	0.34	—	0.32	0.061
1982-83	0.56	0.29	0.32	0.31	0.31	0.32	0.056
1983-84	0.57	0.29	0.32	0.31	0.35	0.34	0.045
1984-85	0.60	0.36	0.39	0.35	0.27	0.38	0.023

Summary and Conclusion

What we thus observed is that—

(i) household finance has consistently moved out from rural and semi-urban areas to urban and metropolitan centres;

(ii) growth rate of co-operative credit is not only less than that of commercial banks'agricultural credit, it is decelerating too;

(iii) commercial banks' credit activity contracts in bad rainfall years; although there seems to be a demand for credit;

(iv) commercial banks have competed with co-operatives, in low risk areas, not complemented them by going to high risks areas;

(v) although *prima facie*, it seems that small farmers have got a larger share of agricultural credit, it is not the actual case;

(vi) co-operatives have more bias to non-small farmers; this bias however is on decline; a closer look reveals that commercial banks too have bias for non-small farmers;

(vii) commercial banks have not penetrated much into the rural areas with agricultural credit; co-operatives on the other hand have contracted; as a result of which total cropped area under short-term credit is more or less constant over time, rather it has declined;

TABLE 3.43: Share of Resources (Per cent) 1984-85

Name of the State	Gross Cropped Area	Short-Term Credit			Long-Term Credit			Total Credit	Deposit
		Commercial Bank	Co-operative Society	Total	Commercial Bank	Land Development Bank	Total		
	(1)	(2)	(3)	(4)	(5)	(6)	(7)	(8)	(9)
Andhra Pradesh	7.6	29.3	7.2	12.7	7.5	14.8	10.4	12.1	8.9
Assam	2.9	0.4	0.1	0.2	0.6	—	0.4	0.2	6.8
Bihar	5.7	1.2	0.9	1.0	3.9	5.0	4.4	1.9	2.0
Gujarat	6.1	4.2	0.4	7.3	7.3	3.5	5.7	6.9	8.2
Karnataka	6.6	9.2	6.3	7.0	11.6	9.2	10.6	7.9	6.9
Kerala	1.7	9.9	16.2	14.6	2.9	6.3	4.3	11.8	3.2
Maharashtra	11.9	5.8	13.6	11.6	7.7	7.2	7.5	10.5	8.5
Madhya Pradesh	13.2	1.3	5.3	4.3	6.9	5.4	6.3	4.9	5.2
Orissa	4.9	1.9	1.9	1.9	1.6	1.2	1.5	1.8	2.0
Punjab	5.7	6.9	16.2	13.9	20.4	24.0	21.9	16.0	15.0
Rajasthan	10.9	1.3	5.0	4.1	5.9	3.2	4.8	4.3	3.8
Tamil Nadu	3.6	22.8	9.4	12.8	10.2	7.2	8.9	11.7	5.9
Uttar Pradesh	14.7	4.3	8.1	7.1	12.4	12.0	12.2	8.5	16.7
West Bengal	4.2	1.5	1.6	1.5	1.2	1.1	1.2	1.4	6.4

TABLE 3.44 : Share of Deposit (June)

Name of the State	1981	1982	1983	1984	1986
Andhra Pradesh	7.21	7.41	7.58	8.93	7.82
Assam	2.29	2.40	2.42	2.41	3.06
Bihar	6.36	6.48	7.12	6.79	8.16
Gujarat	8.91	8.64	8.32	8.23	7.26
Karnataka	7.02	6.85	6.91	6.92	6.37
Kerala	5.23	5.22	3.84	3.15	2.93
Maharashtra	6.18	6.37	6.11	6.22	6.31
Madhya Pradesh	4.47	4.65	4.83	5.22	4.75
Orissa	1.97	2.06	2.17	2.02	2.10
Punjab	15.58	15.98	15.42	14.98	14.79
Rajasthan	3.40	3.46	3.75	3.79	3.68
Tamil Nadu	5.46	5.58	5.76	5.91	6.16
Uttar Pradesh	16.88	16.68	16.99	16.72	17.17
West Bengal	6.61	5.87	6.46	6.37	7.24

(viii) not only that percentage of farmer borrowers is quite small (around 25 %), it is not increasing either. And percentage of non-small farmer borrowers is much more than that of small farmer borrowers;

(ix) recovery rate for small farmers is not necessarily lower than that for non-small farmers although small farmers have lower repayment capacity; recovery rate from non-small farmers was although quite good in earlier years, it is not so in recent years;

(x) branch expansion has helped in deposit mobilisation and deploying long-term credit only which is generally enjoyed by non-small farmers; it has not helped in short-term credit which is more availed by small farmers, and improving recovery rate;

(xi) if need based credit be a criterion, small farmers require a much larger share;

(xii) if agricultural credit be available to small farmers only, a substantial amount of land can be brought under inputs like fertiliser etc., as non-small farmers do not require concessional credit to apply inputs; and this may result a higher agricultural growth because of more input use in totality, and efficient use of resource for agricultural growth by small farmers;

(xiii) the data suggests that agricultural credit has been confined to low risk irrigated areas only;

(xiv) as a result of all these, although regional inequality in branch network of commercial banks has declined in last decade, credit inequality has increased; and

(xv) share of deposit contributed by states is almost constant over time.

All these lead us to conclude that agricultural financial institutions in general and commercial banks in particular—

(1) are risk averse and conservative;

(2) have urban bias and also a bias in favour of non-small farmers; and

(3) take market as given, and not much interested in creating a market as this does not help them in short-run; also this requires bias in favour of rural areas and small farmers which they don't have.

This kind of lending (agriculture) psychology, we feel is not at all conducive for agricultural growth. We require a completely new institution with new lending policy for which a new man is required. Otherwise all cosmetic changes here and there will only help in strengthening the *status quo*.

Notes and References

1. Rath, N. (1987), Dasgupta, R. (1989); Binswanger, Hans P. and Khandekar, S.R. (1989) however observes "effects of credit on crop output is not large despite the fact that credit to agriculture has strongly increased fertiliser use and private investment in machines and livestock".
2. Gadgil, M.V. (1986).
3. Binswanger, H.P. and Khandekar, S.R. op. cit.
4. Rath, N. *op. cit* quoting Fourth Five Year Plan.
5. Desai, D.K. (1988).
6. Gadgil, M.V. *op. cit.*
7. Rath, N., *op. cit.*
8. Patil, R.K. and Dasgupta, R., (1989).
9. NCAER (1975).
10. RBI (1989).
11. Rath, N. *op. cit* quoting Gadgil.
12. Patil, R.K. and Dasgupta, R. *op. cit.*
13. Subba Rao K.G.K. (1990).
14. Rath, N. *op. cit.* (overdue rates have been given).

15. Annual reports of Report on Trend and Progress of Banking in India by Reserve Bank of India. Recovery rates of direct agricultural finance at all India level, according to above reports have improved from 30.2 to 57.1 per cent between 1978 and 1987. Recovery rate of indirect finance is supposed to be 100 per cent.
16. Our estimation is crucially dependent upon the assumption of (i) interest rate and (ii) distribution of disbursement between two seasons. We have in our estimation assumed rate of interest to be 11 per cent, and distribution of disbursement as half and half. With different set of assumptions we obtain following recovery rates for all India for 1986-87 for all farmers together.

r = rate of interest (per cent)

A = Percentage of credit in the beginning of the year (July) i.e. for kharif cultivation.

B = Percentage of credit in the second half of the year (December) i.e. for rabi cultivation.

Recovery Rate, Commercial Bank, All India, 1986+

Interest Rate	Distribution of Credit	A = 25 B = 75	A = 50 B = 50	A = 75 B = 25
$r = 10$		67.6	46.2	29.7
		65.3	44.9	29.1
		65.0	46.8	32.1
		65.9	46.1	30.4
$r = 11$		68.3	46.8*	30.2
		65.9	45.6*	29.7
		65.7	47.4*	32.7
		66.6	46.7*	30.9
$r = 12$		68.9	47.4	30.8
		66.6	46.2	30.2
		66.3	47.9	33.2
		67.3	47.2	31.5

+ In each set 1st row for less than 1 hectare
2nd row for 1-2 hectare
3rd row for more than 2 hectare
4th row for all farmers

* These are the values, we have in our Table 3.26.

17. Binswanger, H.P. and Khandekar, S.R. *op. cit.*

4

ALTERNATIVE APPROACH TO AGRICULTURAL CREDIT

Introduction

What we have thus observed earlier is that banks

(i) are generally risk averse and conservative;
(ii) have urban bias and also a bias in favour of non-small farmers; and
(iii) take market as given; not interested in creating market, and that is why formal agricultural credit has not been able to fulfil its objectives.

The above attitude, however, does not pertain to banking industry alone but is true for Indian economy at large also. What is therefore, required today is a change in the perspective of whole society, not only of banking industry. They cannot be discussed in isolation. However, whereas some changes may be made in banking policy for short term results, it will take a long period to change the attitude of the entire society of which banking sector is an important member. As we can neither afford to ignore short-run considerations, nor agree to the philosophy "we are dead in the long-run", we have to pursue short-run and long-run measures simultaneously. Also it is little easier for us to focus on the reforms of banking industry separately than on the total economic system of the country. We, therefore, confine ourselves to policy issues with regard to banking institutions only.

Short-run Policy

Main problem in banking industry today is the inconsistency in its policies. On the one hand a rural branch manager over time has been given detailed instructions on every aspect of rural lending so that there is hardly any discretion left to him, and is reduced to the position of having to do his work in a mechanical way.[1] On the other hand, the same branch manager has been overnight made the main author of the Service Area Plan (SAP) for the development of village for which he is definitely not ready.[2] Khusroo Committee (henceforth KC) is right to conclude that in agricultural banking policies have often marched ahead of organisation, and organisation ahead of personnel.[3] Even the policies are most of the times not very clear, and lack direction. The so-called salient features of SAP do not throw any light on why the plan? what for the plan? what should be the result? and how should be the performance of the plan or the planner (rural branch manager in this case) be evaluated?[4] In the absence of answer to any such question we will perhaps go back to square one – disbursing the credit and fulfilling financial target but achieving nothing in physical terms. It has, however, been observed that although objectives are not clearly spelt out in the main report "Service Area Approach – Formulation of credit Plan",[5] a set of messages have definitely been given to the branch manager through different levels of hierarchy.[6] The multiple objectives of SAP may then be listed as —

1. increase in production,
2. increase in rural productivity,
3. increase in level of income of rural population,
4. increase in productivity of lending,
5. meaningful lending and effective recycling, (i.e. better recovery is implicit)
6. development of village.

Although above objectives are fair enough, they may conflict among themselves. Whereas first two underline a growth approach, last one explicitly seeks for a developmental approach which may not be synonymous with growth. It is conceded that compared to macro-economic situation where so many conflicts arise in terms of luxury goods *vs* necessary goods, gun *vs* butter, investment goods vs consumption goods etc., agricultural economy poses much less problem. Growth in agricultural income, therefore, seems to be possibility of growth of consumption level. In that case objectives of growth of agricultural production or rural income or more appropriately per capita rural income, and development of village do not seem to be contradictory. Basic tenet seems to be old tricle doen theory which, however, never heppened. Otherwise development is just concerned about income growth or life style of top decile. If development is something different, then the above six objectives of SAA seem to be contradictory unless they are explicitly qualified.

Even in agricultural sector, there is a room for conflict between growth and development although to a lesser extent unless suitable strategies are evolved to achieve both. And it is too much to expect from a rural branch manager who is there perhaps because of his inferior "efficiency level" *vis-a-vis* his other "superior" colleagues who are posted at coveted departments like International banking or Computer division or Advance cells, to formulate developmental strategies. If development is much more than possession of western material products by dominant sections, it is difficult to be incorporated in the performance budget of branches. Other objectives which can be integrated into the performance budgets, however, does not find place there. Although Reserve Bank of India (RBI) in its circular[7] maintains "focus should be on the physical realisation of results in terms of output, productivity, income, employment and not merely on disbursement of credit", only lending for different activities has been considered in the proforma of Block Credit Plan provided in the same circular. Performance budget or SAP not only provides a target, it helps in reviewing the performance also. But what kind of result is expected if the poor branch manager wants to achieve all these six objectives?

Objectives 1 to 4 together, unless qualified, may be viewed as growth approach focusing attention on aggregate measure of per capita product. Because of some psychological barriers, or of class affiliations, as we have observed small borrowers are not considered efficient farmers and reliable borrowers although we have seen earlier that large farmers are no better than small farmers in case of repayment of loans. Irony is that whereas small farmers, today, default because of economic phenomenon of debt trap, large farmers default willingly. Although there is no viable panacea for the latter category, branch managers because of so may socio-political reasons are bound to advance loans to this group. They, however, 'help' the bankers with 'deposits' to fulfil the target of branch managers. Other category i.e. small farmers, on the contrary, can be made to repay the debts. This would, however, take a long time. Small farmers need to be given consumption loan, which may be even for consumer durables and social obligations outside his production activity,[8] and freed from the clutches of money-lenders; their product has to be marketed at a remunerative price; and finally they need to be nurtured such a way that they can resist themselves from going to money lenders again. If this strategy is pursued, even these small farmers may turn out to be the potential depositors; rural sector may become vibrant, and level of savings, most sought after performance indicator in banking industry may go up. This, however, is not so simple as it appears to be. Branch manager has to nurture these small people so that they are not dragged into debt trap of money-lenders again. In the short-run recovery problem which is an important objective in the earlier list may worsen as substantial amount of loan to small people has been given for consumption purpose. Also "deposit" target in the short-run may not be fulfilled by branch managers as large farmers no longer would like to oblige them by keeping their deposit in the bank for "few days". Branch Manager therefore has to wait for a long period what he perhaps cannot afford under present circumstances, as his performance is judged by short-run outcomes. If objectives 1 to 4 along with deposit level therefore become important items of performance budget, branch manager has no option other then channellising the fund to enterprising and productive farmers with well endowed plots. And that's why the objectives of production, productivity, employment etc., inspite of being mentioned in circulars do not find places in implementation mechanisms.

Credit Policy

It therefore, seems that the vested interest at the decision-making level, supposedly dictated by dominant groups, is one of the reasons for myopic attitude of banking industry. One important aspect of the new approach is to completely change the emphasis. In to-day's context, the whole banking system is obsessed with financial targets, as if that is the end of everything. The targets in fact have to be in physical terms. Objective 1 to 3 have to be converted into physical targets, albeit with modification. Unqualified statement of increase in production or rural income cannot be valid objectives. If this policy is meant for smaller people, the objective has to be increase in production by small farmers or increase in income of poorer group. Otherwise credit will flow to richer people to increase the inequality.

The credit policy has therefore, to be more specific, and performance has to be measured in terms of physical output, not financial target. Productivity of lending can then be calculated as —

$$Pr_t = AGWI/TCD_{t-1} = (\sum_{i=1}^{L} AGI_{it} * W_i)/TCD_{t-1}$$

where Pr_t = Productivity of lending in the year *t*

$AGWI$ = Additional generated weighted income

AGI_i = Additional income generated in the *i*th income class

W_i = Weights assigned to ith income group

L = Number of income classes

TCD_{t-1} = Total credit deployed in the area in previous year

As estimation of class-wise agricultural production, not to speak of agricultural income, for every year may be a difficult task for the branch managers, productivity has to be calculated in a different fashion. Deposit instead of income can be considered as an indicator. Poorer people, however, may resort to consumption loan which will not yield income in that short-run. Also because of so many reasons, they may not generate enough savings for deposit. Deposit therefore has to be weighted much heavily in favour of poorer people. In this case the branch manager will be inclined to mobilise deposits more from smaller people. Productivity of lending can then be calculated as—

$$PrD_t = AGWD/TCD_{t-1} = \sum_{i=1}^{L} (AGD_{it} * W_t)/TCD_{t-1}$$

where

PrD_t = Productivity of lending - deposit approach

$AGWD$ = Aggregate weighted average of deposits belonging to debtor to the branch/bank

AGD_{it} = Minimum level of deposit on a day during a predetermined period of at least three months; belonging to debtors of the bank/branch

W_i = Weights assigned to each income class ($\Sigma W_i = 1$)

TCD_{t-1} = Total credit deployed in the previous year.

Assignment of W_i to different income classes should be an integral part of credit policy. Weights should be so heavily loaded in favour of smaller people, so that branch manager does his best to get deposit from them. Since the branch manager will not have any legal powers to extract these deposits, he will have to establish a personal relationship with his smaller clients, and see that their income rises through credit. This deposit may later be adjusted with recovery amount.

Similarly productivity of lending can be measured in terms of employment generation or foodgrains production or reduction of inequality etc.

Recovery rate as a performance indicator on the other hand has to be calculated with a completely different emphasis since large farmers have been considered as willful defaulters, and branch manager should be given an incentive for recovering the due amount from large farmers. Recovery rate as performance indicator (RRP) can therefore be estimated as given below

$$RRP_t = \sum_{i=1}^{L} AR_{it} * g_i / TD_t$$

RRP_t = Recovery rate as performance indicator in year t
AR_{it} = Amount recovered from ith income class in year t
g_i = Weight assigned to ith income class in year t ($\Sigma g_i = 1$)
TD_t = total demand raised in year t
L = Number of income classes

Since branch manager is made to give more attention towards recovery from large farmers, g_i enunciated in credit policy has to be loaded quite heavily towards large farmers. These two indicators *PrD* and *RRP* together are supposed to bring results as former will make the branch manager to increase the income and hence deposit of smaller farmers and latter will force them to recover dues from large farmers.

Lending Methodology

The above policy, however, can work meaningfully in a particular circumstance only. This may require a new lending methodology which primarily asks for an organic relationship between borrowers and banks; organising the people; and people's participation in the lending decisions. People, here of course is not represented by elites of villages, but by masses.

This new methodology is supposed to have following salient features.[9]

(i) *Team Work*

It is not the branch manager alone, but the whole branch staff under leadership of branch manager that take part in this lending process. The first task of the branch manager is therefore to take initiative in the team building, and give leadership.

(ii) *Integration*

Integration i.e. establishing an organic link between bank and village is the first task of the branch team. The team has to understand the functioning of the village during this phase. Branch manager and its team must have detailed knowledge of each and every household, not of few of the villages. It is necessary for them to analyse the various dimensions of the village – production economy, marketing system, income-consumption balance of households, employment opportunity, peak and lean season, problems, potential of people and their aspirations, exploitative elements operating in the village etc. The branch has to come closer to the people during this phase and prove themselves to be friend, philosopher and guide of the people. This is in that sense an investment phase from which return would come much later. The branch may take different steps within some broad guidelines to establish this organic link between bank and people.

(iii) *Organising Poorer People*

Next step of the team is to bring the people together, make them self confident, help in understanding economic process, instill co-operative spirit to work in a group, educate them about group economics, inculcate concern for others, and also enlighten them about their rights. It cannot be denied that to-day whoever, whether agricultural or industrial classes or white collar people like academicians, bureaucrats and bankers or blue collar union workers, have got their shares is because of their organisational strength and bargaining power. People in the village too therefore need to be organised so that (i) they can demand their share and more importantly (ii) work together with a co-operative spirit for economies of scale in procurement, production and marketing. This group needs to be different from trade unions so far as vested interest is concerned. It is in fact this co-operative spirit which has brought tremendous success even in competitive societies like Japan, South Korea, Netherlands and West Germany.[10] These can be achieved by making forums, groups, village councils etc.

(iv) *Formulation of People's Action Plan (PAP)*

After knowing the village, and organising the people into homogeneous cohesive groups, the branch team has to initiate the planning process with the help of people. People have to be told about the objectives of the plan action first. People must take part in credit planning with listing out their priorities, selection of beneficiaries, discussion of problems etc. KC[11] rightly observes that SAA should be graduated to development of Area Approach, and economic development planning is meaningless without economic and organisational content and an input of popular wishes and aspirations. Branch manager, however, has to consider the economics of those activities, and constraints of the branch. Regional manager needs to advise the branch manager in plan making.

After the plan has been prepared in the branch, it has to be discussed with the people, and they need to be explained about the realities of the proposed credit plan. It is worthwhile if the document be written in local language. Group principle of Development of Women and Children in Rural Areas (DWCRA) may be applied whenever possible.

(v) *Implementation and Monitoring*

Next task of the branch team and the people is to see that plan be implemented with all seriousness. All the activities then have to be monitored, and problems to be solved by the people and branch team together so that objectives of the plan are achieved. Initially, the branch team may have to put more efforts. But gradually the people can take over this job, and monitor the utilisation of credit along with solving their problems with the help of bank and other government agencies more efficiently.

While it is conceded that this approach cannot be implemented in all six lakh villages over night, this should not be ruled out as impossible either. National Institute of Bank Management while pursuing experiments in this line since 1983 has already got some worthwhile results.[12] At one branch at Kerala the bank had taken up the issue of alienation of land of tribals to the court, and freed the tribals from money-lenders' bondage. In another branch at Maharashtra the bank and people together made 'bunds' and solved the water problem. Drinking water problem was solved in a village in Andhra Pradesh by laying pipes to bring water from a nearby stream. Interesting point, however, was that scheduled class and other poorer people were so conscious that they contributed for this at par with richer landlords so that the latter do not corner more water.

These are not one or two stray cases, as experiment now is going on in about 500 rural branches. One thing, however, is certain that whenever people's participation has been obtained branch's performance in terms of recovery, deposit mobilisation, housekeeping etc. has improved.

Reward System

What we find above is that new credit policy along with it's lending methodology puts tremendous pressure on branch manager, and his job becomes extremely challenging. Unless there is sufficient reward there will definitely be scarcity of personnel to carry out the above credit policy with different lending methodology.

The major requirement in this framework is to change the attitude towards rural banking. Whereas agricultural credit is a part of priority sector, personnel policies are least conducive for rural postings. Not only that staff at rural branches have to work under several infrastructural constraints, they are treated as second class staffs. Whereas a young bright blue eyed officer is supposed to bag a posting at international branch or computer division or merchant banking, others are sent for rural postings. Not only that they face problems with health, education etc., they at rural areas seem to lose in monetary terms also. Objective of setting up Regional Rural Banks (RRBs) was to reduce the cost of rural banking by paying less to RRB staffs.[13] KC,[14] however, now has rightly recommended merging of RRBs with sponsor banks. But damage has already been done.

The first correction to be made therefore is to give importance to rural postings. This should not be a pre-condition for promotion. Rural postings on the other hand has to be reserved for the better staffs. depending on their performance, as discussed earlier, they can later be eligible for some non-material benefits. While serving at rural areas they need to be compensated monetarily so that they can take care of their children's education, and such other items elsewhere till the facilities in rural areas are available. As these types of problems are much less for younger people, and they possess much more of energy and dynamism, younger generation of officers below the age of thirty five should be in charge of rural branches where they can innovate.

Organisational Change

Since commercial banks' prime interest is in recovering their debt for recycling, banks' job is much beyond the disbursement of credit and even ensuring production. Bank should ensure marketing of the products at remunerative price also. This is more true for non-farm products. It would, however, be impossible if not inefficient for banks to specialise in marketing also. It can either open its own specialised subsidiaries for marketing or make tie-up arrangements between un-exploitative marketing agencies and village producers.

As in DWCRA[15] model of Velhe, informal production groups at village level and registered co-operative society at district level taking care of input and output can be considered. As suggested by KC one block should be assigned to one commercial bank which should be solely responsible for the development of that area. Since all nationalised banks in the country work under same monetary policy and RBI guidelines, competition among the commercial banks is a rather restrictive concept in this country. Also, competition among banks will definitely be not for catering to weaker section – but for wooing the richer. As observed by KC[16] this will not only reduce operational cost, and obtain scale economies, but have external effects also. The bank being dominantly present in a block (instead of so many banks operating) will strive for social amenities like school, health, clinics, hospitals at a central point in block.

The ideal case will, however, be to separate the banking activities of rural and semi urban (RSU) areas from those of urban and metropolitan (UM) areas. Whereas the latter institution may be privatised, the former one in RSU areas must be in the public sector. Co-operatives and RRBs too may be merged with this new RSU bank. Each state may have its own RSU bank. This will help in preventing the flight of houschold capital from backward to developed areas; or from RSU to UM areas. Performance criteria of these RSU banks then need to be shifted from financial to physical indicators as discussed earlier. Monopoly and accountability rather than competition have to be guiding factors for these banks. A new priority sector then can be defined in terms of small farmers, unirrigated areas, cereal crops, weak entrepreneurs, etc. Fifty per cent of credit may be enmarked for this category. A subsidiary of this RSU bank can then take care of forward-backward linkages, and other developmental aspects.

As agricultural sector can not absorb the total rural surplus labour; nonfarm sector has to be developed. However, programmes like IRDP, SEPUP, SEEUY can not succeed as economic strategies. They can at best be perceived as soft welfare measures. Entrepreneurship is not so easy. Backward, illiterate mass can at best acquire skills and expect wages. All forward and backward linkages need to be provided by a specialised organisation. Even in China, where rural industrialisation has been so successful; the term township village and private enterprise (TVP) has become a magic word; literacy is very high; entrepreneurship potential is quite abundent; more than ninety per cent of the workers do not have any desire to run an enterprise there.[17]

This new subsidiary of RSU is to provide raw materials, technologies and guidance to small producation groups at RSU areas, and market their products to national and international market. This organisation must be a part of the total production system, enjoying the profit and bearing the losses. Employees in these organisations may have a low basic wage and high dividend components in their remuneration packages. Piece rate wage compared to time rate may be given more importance for the tiny producers who now work as wage earners. They must have, however, share in the profits through dividends. This developmental organisations may even take up the responsibilities of local level government departments providing infrastructural and developmental inputs so that problem of coordination among different agencies is reduced. (Chart-1)

Long-Run Dimensions

All said and done, the new credit policy and lending methodology for the development of mass can work only if the attitude of the delivery system is conducive for this purpose.

According to KC,[18] the relative lack of adequately trained personnel has been a major constraint in both the deepening and widening of credit in pursuit of rural development. The Report further observes "The problems of commercial banks in the development of banking for rural lending are low motivation as well as gaps both in knowledge and skills. Above all, the low level of sympathy for the disadvantaged classes and over shadowing all these, the bureaucratic approach are the other causes".

Although skill and knowledge are not strictly speaking value neutral, they can yet be imparted through regular training courses to bank staffs. Proper attitude and culture, sympathy for disadvantaged classes, concern for others etc. are, however, by-products of values which are formed over a long period. Society is responsible for imparting these traits. But since banks have already a stock of staff members with different value orientations, and the society cannot inject new values to these banking personnel, banks have to take initiative to inculcate proper values and attitudes in their staff members so that they can utilise knowledge and skill meaningfully for the development of the society.

Importance of values and theoretical understanding along with knowledge of skills in achieving organisational goals may be explained through Chart 2. All organisations need to realise that they owe their origin to the society; are accountable to the society; and are to contribute to the society.

Towards a New Value

On the one hand Weber's assertion that "ethic is given? and "emergence of an economy compatible with this ethic is inevitable" does not seem to be correct as Japan and China both having similar ethics of Confucianism have adopted different types of economy, and capitalistic traits have been found to be available in non-Protestant communities like Gujaratis, Parsees in India. On the other hand, Marx's contention that ideology and ethics are "no more than reflections of underlying material conditions – in particular economic conditions"[19] also do not seem to be valid. Whereas cut throat competition and individualism are the cultural dimensions in US economy, Swedish material conditions have not affected the co-operative spirit there. A system of value perhaps may be generated.

Theoretical Input

However, what is required for all the bankers first is the understanding of workings of an economy : the importance of agriculture in the economic growth process. Agricultural products constitute the most important segment of wage basket. Agricultural bottleneck therefore causes inflation, reduces profitability of industrial sector, and may turn to be the bottleneck for inducement to investment in industrial sector, and hence agricultural bottleneck may prove to be barrier to economic growth. Agriculture, however, not only provides food and raw material, it provides a market for industrial goods. It would be much profitable for industrialists if the size of the market is big not less; if the market constitutes not only of large and big farmers but of small farmers and landless labourers. Healthy long-run growth may not take

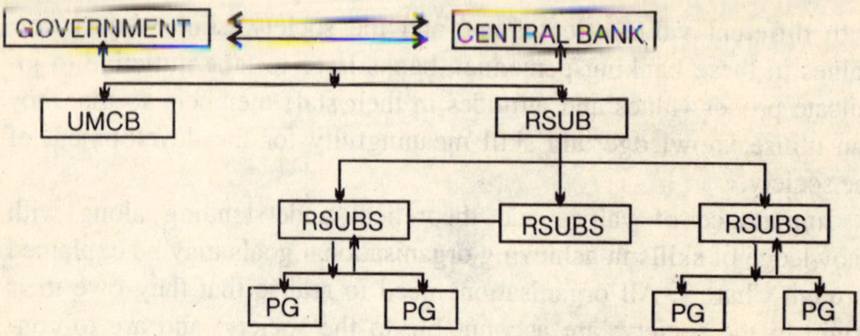

UMCB : Urban and Metropolitan Commercial Bank : Operating only in urban and metropolitan areas.
RSUB : Rural and Semi-Urban Bank operating in rural and semi-urban areas, doing both commercial and development banking.
RSUBS : Rural and Semi-Urban Bank Subsidiary : Taking care of non-banking and developmental aspects including backward and forward linkages.
PG : Production Group of weaker sections in rural and semi-urban areas.
→ : Direction of Control/ Influence.

CHART 1

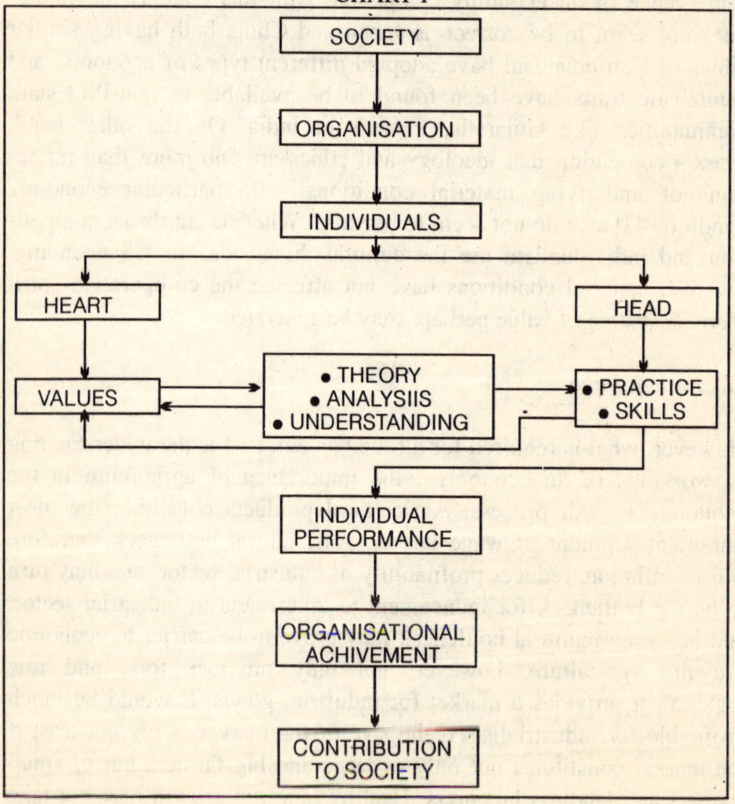

CHART 2

place in the absence of such market. Economy may stagnate or even collapse if such market is not generated. Supply may create its own demand for the time being, but not for a long period. In the long-run economy can grow only if demand position improves. In other words attitude of supply creates its own demands has to be transformed to "demand creates its own supply".[20] If our bankers are convinced about this proposition, then only they may consider whole range of priority sector lending specially to smaller people as commercial proposition, and not a charitable act. And moment they consider them commercial proposition, they may take pride in this type of lending (provided their value system is correct), and quality of lending may improve (ignoring the fact that quality of industrial lending is not much better). Understanding the importance of market, however, is not enough. As development of this broad-based market takes time one has to take a long-run attitude. This would only be possible if they have understanding of Schumpetarian entrepreneurship.[21] They need to give priority to autonomous rather than induced investment; appreciate that even the consumption loan to smaller people is in fact an investment loan in true sense; and hence take the risk and have patience to wait for a longer period for the return.

Even this indoctrination, however, may not be sufficient for meaningful implementation of the credit policy with new lending methodology if their internal values do not conform to above framework. As shown in the diagram even the theoretical position of an individual is influenced by his internal values.

Values

It is not essential to turn to other religion or culture for finding values for us. Unfortunately, we have to resort to that in India because we do not have a national value today. Our value system has been completely destroyed by the colonial domination of more than three hundred years. One option is to dig our original values and propagate them; and if necessary impose through the curriculums of school, college, university and training centres. Other option is to analyse the value of other society, study its impact on that economy, examine adaptability to Indian condition, and finally come out with desired set of values to be spread. As the former runs the risk of not having being tried, the latter seems to be a better option. Study of Japanese Society, in that context seems to be more prudent because of so many reasons. Japanese values may prove to be less alien, if not more familiar to Indian society since ("Throughout Japanese history up to the present individualism has never prospered"[22]) our culture too if not destroyed was to have taught us the principle of "Vasudaiva Kutumbakam" i.e. anti individualism. Japanese Confucian ethics did not have much conflict with the religion of Buddhism which owes its origin to Indian society. Japanese economic thinking conforms to our theoretical framework enunciated earlier. They have learnt to take long-term view. It is reflected even in their to-day's stock market system where major portion of share is held by banks which are not bothered for short-term dividends. Matsushita's mission "an inexhaustible supply of goods"[23] is nothing but a song of large market. More importantly neo-classicism although took a root, it failed to attain the overwhelming ascendancy as in other English speaking world; and Japanese economic theory could become innovative because of influence of thinking of two equally significant schools a 'modern path' which is a synthesis of neo-classical and Keynesian ideas, and the other a somewhat more isolated and inward looking Marxian path.[24]

Economic Thoughts

Assimilation of extreme ideas of neo-classical and marxian tradition was possible only because of their value system. Right in 17th century in Japan, poverty had been identified as an inescapable consequence of luxury, and to Kumazaw Banzan (1619-91) "the solution to the nation's economic woes lay not in extending the area of agriculture, but in a more fundamental reform of the whole system"; his policy, rested on 'revering grain and despising gold'.[25] According to Sato Nobuhiroh (1769-1850) 'benevolent ruler's greatest duty is to relieve the sufferings of the people'[26] Nobuhiro's wise government not only practices the traditional values of charity and justice, but one which actively intervenes to promote prosperity through the introduction of new techniques and new types of production suggesting an important role of State in the development process.[27] This is reflected in even today's state of affairs when in spite of substantial western influence state plays an important role even in economic decisions in Japan. Nobuhiro had laid great emphasis on education, and proposed for a national school system to provide a basic moral education for all children from the age of eight.[28] Although there were other thinkers like Kaiho Seiryo, Kanda Takahira who were vowing for free market system and individualism, general concensus drifted increasingly towards nationalist and conservative philosophies. Nobuhiro's emphasis on education had larger impacts on later national policies. German ideologies became more important Ricardo's comparative advantage theory was attacked and a sense of nationalism developed. Following Freidrich List, Japan wanted to have its own economic principles, as social and economic principle were thought to be time and place specific. Japan was also influenced by Gustav Von Schmoller, another German economist who strongly felt that the study of the economy was not morally neutral positivist science, but a means to the achievement of a healthy and harmoneous society.

In fact social policy school's approach which was very much influenced by Schmoller's concern of understanding the economic causes of social problems such as poverty and class conflict and devising policies to ameliorate these problems corresponded very well to the pre Meiji Japanese tradition of Keikoku Saimin :'administering the nation and relieving the sufferings of the people'.[29] According to Kawakami, the ultimate object of economics is to make human beings more fully human, not merely to study the production and distribution of wealth.[30] The Association for social policy while rejecting socialism opposed *laissez faire* also because the extreme exercise of self interest and uncontrolled free competition would result in an excessive gulf between the rich and the poor.[31] It advocated the economic system with a private property with both individuals and state taking part in preventing class conflict and promoting social harmony. State had further been seen as an organic body in which the interests of the individual were submerged, and whose harmonious growth is the supreme goal.[32] Even neo-classicical economist like Uzawa[33] in recent period feels that "Socialism characterised as it was by co-operation, equality and the enhancement of culture, indeed represented the direction in which Japan should advance".

What we therefore find that Japanese economic thought since long has been moulded by the concern of welfare of human beings. They did not yield to any particular type of economics, but developed a theory of economics desirable for their country. Basic elements of this seem to be (i) nationalism, (ii) importance of state in economic decisions, (iii) long-run attitude, and (iv) well being of all the fellow Japanese and harmonious growth. All these had been backed up by their education system which had a strong component of moral education which was guided by a particular ethics – Confucianism. Influence of all this can perhaps be best understood by the philosophy of business of Matsushita "The mission of a manufacturer is to overcome poverty, to relieve Society as a whole from the misery of poverty and bring wealth. Business and production are not meant simply to enrich the shop or the factories of the enterprise concerned, but all of society".[34]

Confucian Elements

In a Confucian society person's worth is estimated not on the basis of whether or not he possess money, but on the basis of whether or not he has been educated.[35] And the education in Japan had been geared not to the individual, but largely to the purpose of that state. The first clause of the Imperial Universities ordinance clearly states 'The aim of the imperial universities shall be to teach the arts and science in accordance with the priorities of the state, and to make a study of their principles. Even other educational institutions had to produce people whose talents would be of use to the nation. Importance of education in public service can be gauged from Nobuhiro's Memorandum which is foundation of education policy of later Japanese governments. It says : "Male children who display particular intelligence are to be selected for a more rigorous education which culminates in training for public service. The educational authorities of the Utopian state are also to be in charge of a comprehensive network of welfare institutions, including hospitals and nurseries for the children of the poor".[36] All these clearly show how a particular value system was imbibed into the Japanese community by educational system. What is felt here is that a new ethics can now also be generated by constant effort through the medium of education over a quite long period of time.

Knowledge about Confucianism therefore becomes important for two reasons (i) even if it is alien ethics, it can be injected, if it is found desirable and (ii) it is not very alien to Indian culture. Essence of Confucian teaching, after all is "the goal for the individual is the development of personality until the ideal of a perfect man, a true gentleman, a sage is reached. On the other hand, his goal for society is universal order and harmony under the rule of a perfect sage".[37]

Confucius regarded[38] Benevolence (Sen), Justice (i), Ceremony (*li*), Knowledge (*chih*) and Faith (*hsin*) as among the most important virtues, but believed that of these it is benevolence (sen) which was the virtue which must be at the heart of humanity. Benevolence again has to be tempered by justice and reinforced by knowledge; a simple spontaneous humanity is not enough". In addition harmony signifying people's being in accord with one another and preserving accord within society was essential for the achievement of benevolence.[39] It is thus clear that head and heart are very much together in confuscian thinking. Knowledge and humanity/justice constitute benevolence which is the basis of above thinking. And harmony signifying people's being in accord with one another and preserving accord within society is most essential for the achievement of benevolence.

Only those who acquire this kind of perfect love of humanity are spoken of as men of virtue. And for a perfect gentleman there are nine following considerations:[40]

(i) desire to see clearly when he looks something;
(ii) desire to hear every detail when listening to something;
(iii) desire to present a tranquil countenance;
(iv) desire to preserve an attitude of respect;
(v) desire to be sincere in his words;
(vi) desire to be careful in his works;
(vii) willingness to enquire further into anything about which he has doubts;
(viii) willingness to bear in mind the difficulties consequent on anger; and
(ix) willingness to consider moral values when presented with the possibility of profit.

While others help to synthesise different view points, the last one gives a new meaning to profit which unlike in western thinking, cannot be value free. That is why confucius' maxim is 'guidance by morality, control by ceremony' which underlines that 'people will come to have a sense of moral shame, and to act correctly'.[41] As a result unlike in America money making was in no way a sufficient condition for achieving the respect of society . . . money making was often even a condition for social censure.[42] This morality reflected through virtues of harmony, propriety, loyalty, benevolance, sincerity and righteousness had been highly advocated in the constitution modelled by Shotoku Taishi way back in seventh century A.D. Even few centuries later when Meiji revolution was laying the foundation for the building of a modern state on the western model, Japanese seem to have fully rejected western spiritual ideas.[43] This was possible because Confucianism had spread widely and deeply among the Japanese people.

As the nine considerations of confucianism advocate respect towards all, and develop the attitude of analysing and synthesis to get the best, Japan in Meiji period followed and adapted French educational system, British Navy, French army, British telegraph and railway system, American University patterns, German Civil code and French criminal code.[44]

All these confucian elements were perhaps more pronounced in pre-war education system, as education during that period was geared not to the individual, but largely to the purpose of state.[45] However, even in post war "San Fransico" regime when people in general seem to have lost confidence in traditional culture and life style, western individualism and liberalism have not yet become firmly rooted in Japan.[46]

Task Ahead

What we therefore find is that factors behind Japan's success story are policy directions, economic theories, economic thoughts and values which are highly interrelated, and very different from dominant western thinking. There is therefore a strong case in India also to adapt a set of value, culture, economic thought, economic theory and policy which is more meaningful to us and also less alien to us. Confucian philosophy and Japanese economic principle may be a good foundation for this purpose.

It is indeed a long-term phenomenon. Also, it is mainly the duty of the society or the state to formulate such educational policy, design appropriate educational system, and finally mould the society with desirable value and culture so that banks and other organisations can recruit appropriate persons to fulfil their objectives. Banks, however, have to take lead in between. They will be able to meaningfully translate new credit policy and lending methodology into realities of (i) upliftment of poor, and (ii) ushering of phase of stable growth only if they can internalise the above economic and social values through their own training institutions.

They therefore need to look into their training system including the training inputs, and make radical changes. Training has to be the "priority sector" of banking industry till the entire organisation acquires necessary skills, understands correct economic theories, and imbibes new values underlining (i) nationalism, (ii) importance of state in economic decisions, (iii) long-run attitude, and (iv) concern for welfare of fellow countrymen and harmonious growth to achieve the goal for which it had been created and hence justifies it existence.

Notes and References

1. Agriculture Credit Review Committee, (henceforth KC), 1989, p 126
2. Dasgupta, R., (1989).
3. KC *op. cit*, p 490.
4. Dasgupta. R. *op. cit.*
5. Rural Planning and Credit Department (RPCD), (1988).
6. Dasgupta, R. *op. cit.*
7. RPCD, (1988).
8. KC *op. cit.*, p 256.
9. Rao, K.D. *et al*, (1987).
10. KC *op. cit*, p 168.
11. KC *op. cit*, pp 426-430
12. Rao, K.D. *et al op. cit*. also see Aadhar : PDV Special Issue, 1992.
13. KC *op. cit*, p 131.
14. *ibid.*, pp 152-153.
15. Rao, K.D.R., *et al*, (1990).
16. KC *op. cit*. p 431
17. Byrd, A William and Quingsong (1990), p 286.
18. KC *op.cit*. p 490
19. Morishima, Micheo (1984), p 1.
20. Rakshit, M. (1983), pp 3-5.
21. Schumpeter, Joseph, (1934).
22. Morishma, M., *op. cit*. p 8
23. Matsushita Konosuke (1984), p 22.
24. Morris Suzuki, T.M., (1989).
25. *ibid.*, p 15.
26. *ibid.*, p 35.
27. *ibid.*, p 35.
28. *ibid.*, p 36.
29. *ibid.*, p 63.
30. *ibid.*, p 79.
31. *ibid.*, p 64.
32. *ibid.*, p 66-67.
33. *ibid.*, p 103.
34. Matsushita, K., *op. cit.*, p 22.
35. Morishima, M., *op cit*, p 178
36. Morris Suzuki, T. *op. cit.*, p 36
37. Smith, D.H. (1974), p 62.
38. Morishma, M. *op. cit.*, p 3.
39. *ibid.*, p 4.
40. *ibid.*, pp 4-5.
41. *ibid.*, p 5.
42. *ibid.*, p 91.
43. *ibid.*, p 52.
44. *ibid.*, pp 88-89.
45. *ibid.*, p 175.
46. *ibid.*, p 169.

BIBLIOGRAPHY

Agriculture Credit Review Committee, *A Review of The Agricultural Credit System in India*, (Chairman Khusro, AM), Reserve Bank of India, Bombay, 1989.

Bandyopadhyay, Arun K., *Economics of Agricultural Credit*, Agricole Publishing Academy, New Delhi, 1984.

Bardhan, Pranab, *Land, Labour and Rural Poverty:Essays in Development Economics*, Oxford University Press, Delhi, 1984.

Bardhan, Pranab and Rudra, Ashok, "Terms of Sharecropping contracts", *The Journal of Development Studies*, April 1980

Basu, Kaushik, *The Less Developed Economy: A Critique of Contemporary Theory*, Oxford University Press, Delhi, 1984.

Bhaduri, Amit, "A Study in Agricultural Backwardness Under Semifeudalism", *The Economic Journal*, March 1973.

Bhaduri, Amit, *The Economic Structure of Backward Agriculture*, Macmillan India Ltd, Delhi, 1984.

Binswanger, H.P. and Khandekar, S.R., "Determinants and Effects of the Expansion of the Financial System in Rural India" (Mimeo) Agricultural Operations Department II, Latin America and the Caribbean Region, World Bank, Washington D.C., 1989.

Byard, A. William and Qingsong, L. (eds), *China's Rural Industry : Structure Development and Reform,* Oxford University Press, (for the World Bank), Washington D.C., 1990

Central Statistical Organisation, *Statistical Abstract in Organisation India*, Government of India, New Delhi (Annual Reports).

Cramer, J S, *Econometric Applications of Maximum Likelihood Methods*, Cambridge University Press, Cambridge, 1986.

Dasgupta, R, *Nutritional Planning in India*, Studies in Economic Development and Planning No. 45, Institute of Economic Growth, Navarang, New Delhi, 1989

Dasgupta, R, *Analysis of Agricultural Performance without any Analytical Framework 1952-82*, (Working Paper), NIBM, Pune, 1986.

Dasgupta, R, *Determinants of Agricultural Performance*, Report, National Institute of Bank Management, Pune, 1989.

Dasgupta, R, "Contradictions in Service Area Approach", paper invited for discussion in the Seminar on Data Base for Poverty Studies, The Indian Econometric Society, Ahmedabad, 1989.

Dasgupta, R, *Statistical Analysis of Agricultural Credit*, Report, National Institute of Bank Management, Pune, 1990.

Dasgupta, R, "Alternative Approach to Social Banking", paper presented to 15th Bank Economists' Conference, Baroda, 1992.

Dasgupta, R, "Narasimham Committee Report" : A Critique", *Business Herald*, Pune, February 13, 1992.

Dasgupta, R, "Financial Reform for Rural and Semi Urban Areas : Pursue or Perish", *Financial Express*, Bombay, April 16, 1992.

Department of Agricultural Cooperative, *Area and Production of Principle Crops*, Government of India, New Delhi (Annual Reports).

Department of Agriculture, *All India Report on Agricultural Census, 1970-71*, Ministry of Agriculture and Irrigation, Government of India, New Delhi, 1975.

Department of Agriculture and co-operation, *All India Report on Agricultural Census 1980-81*, Ministry of Agriculture, Government of India, New Delhi, 1987.

Desai, D K, "Institutional Credit Requirements for Agricultural Production - 2000 A.D., *Indian Journal of Agricultural Economics*, July-September, 1988.

Fomby, Thomas B, Hill, R Carter, and Johnson, Stanely, *Advanced Econometric Methods*, R. Springer-Verlag, New York, 1984.

Gadgil, M V, "Agricultural Credit in India : A Review of Performance and Policies", *Indian Journal of Agricultural Economics*, July- September, 1986.

Ghose, A K and Saith, A, "Indebtedness, Tenancy and the Adoption of New Technology in Semi-Feudal Agriculture", *World Development*, April 1976

Gokhale Institute of Politics and Economics, *Writings and Speeches of Professor D R Gadgil on Co-operation*, Poona, 1975.

Hirway, Indira, *Abolition of Poverty in India*, Vikas Publishing House, New Delhi, 1986.

Hossain, Mahabub, "Agrarian Reform in Asia – A Review of Recent Experiences in Selected Countries" in *Rural Poverty and Agrarian Reform*, (eds.) Jones, S., Joshi, P.C., and Murmis, M., Allied Publishers, New Delhi 1982.

Johnston, J, *Econometric Methods*, McGraw Hill Kongakushe, New York/Tokyo, 1972.

Krishnaji, N, "Measuring Agricultural Growth", *Indian Journal of Agricultural Economics*, Vol 35, 1980.

Kurulkar, R P, *Rural Employment Guarantee Scheme and Landless Agricultural Labour*, Himalaya Publishing House, Bombay, 1989.

Matsushita Konosuke, *Not for bread Alone*, PHP Institute InC., Tokyo, 1984.

Mitra, Ashok, "Bumper Harvest has created some dangerous illusion" *The Statesman* 14-15 October, 1968.

Morishima, Micheo, *Why has Japan Succeeded?*, Cambridge, London, 1984.

Morris Suzuki, Tessa, *A History of Japanese Economic Thought*, Roultledge, London, 1989.

National Bank for Agriculture and Rural Development (NABARD), *Statistical Statements Relating to the Cooperative Movement in India*, Bombay, (Annual Reports).

National Council of Applied Economic Research (NCAER), *Changes in Rural Income in India, 1968-69, 1969-70, 1970-71,* New Delhi, 1975.

Newbery, D G M, "Tenurial Obstacles to Innovation", *The Journal of Development Studies*, July 1975.

Oriental Bank of Commerce, *Aadhar : PDV Special Issue, Vol. 11,* No. 2, March-June 92, Delhi.

Patil, R K and Dasgupta, R, *Evaluation Study of Girna Irrigation Project*, Report, National Institute of Bank Management, Pune, 1989.

Patnaik, Utsa, *The Agrarian Question and the Development of Capitalism in India*, Oxford University Press, Delhi, 1986.

Prasad, Pradhan H, *Lopsided Growth Political Economy of Indian Development*, Oxford University Press, Delhi, 1989.

Raj, K N, "Mobilization of the Rural Economy and the Asian Experience" in *The State of Development Economics, Progress and Perspective*, (eds) Ranis, G. and Schultz, T Paul, Basil Blackwell, Oxford,1988.

Rath, N, "Agricultural Growth, Investment and Credit in India", Sir A Ramaswami Mudaliar Endowment Lecture, University of Kerala, Trivandrum, 1987.

Rao, K D R *et al*, *Planning for Development of Village*, Report, NIBM, Pune, 1987.

Rao, K D R, *et al, Development of Women and Children in Rural Area (DWCRA),* Velhe District, Pune Report, NIBM, Pune, 1990.

Ray, S K, "Instability in Indian Agriculture Revisited" in *Recent Advances in Agricultural Statistics Research* (eds) Prem Narain *et al.* Wiley Eastern Limited, New Delhi, 1991.

Reserve Bank of India, *Banking Statistics* : *Basic Returns*, Bombay, (Annual Reports).

Reserve Bank of India, *Report on Currency and Finance,* Vol II, Statistical Statements, Bombay (Annual Reports).

Reserve Bank of India, *Report on Trend and Progress of Banking in India*, Bombay, (Annual Reports).

Reserve Bank of India, *Statistical Tables Relating to Banks in India*, Bombay, (Annual Reports).

Reserve Bank of India. *All India Debt and Investment Survey 1981-82* : *Statistical Tables Relating to Cash Borrowings and Repayments of Households during July 1981 to June 1982 and Cashdues outstanding as on 30th June, 1982*, Bombay, 1989.

Rudra, A, *Indian Agricultural Economics Myths and Realities*, Allied Publishers Private Ltd., New Delhi, 1982.

Rural Planning and Credit Department (RPCD), *Service Area Approach – Formulation of Credit Plan*, RBI, Bombay, 1988.

Rural Planning and Credit Department (RPCD), *Service Area Approach – Preparation of Credit Plan,* Circular No RPCD, L BS BC, 28/65 -88/89, RBI, Bombay, 1988.

Schumpeter, Joseph, *The Theory of Economic Development*, M A, Cambridge, 1934.

Subba Rao, K G K, "Cost of Debt of Household Sector : A Note", *Economic and Political Weekly*, July 7, 1990.

Smith, D H, *Confucius*, Paladin, London, 1974,

The Fertilizer Association of India (Annual Reports). *Fertilizer Statistics*, New Delhi,

Thorner, Daniel and Thorner, Alice, *Land and Labour in India*, Asia Publishing House, Bombay, 1962.

Vaidyanathan, A, "Constraints on Growth and Policy Options", *Economic and Political Weekly*, September 17, 1977

Vaidyanathan, A, *India's Agricultural Development in a Regional Perspective*, Centre for Studies in Social Science, Calcutta, 1988.

INDEX